T0146238

Instrumental Intimacy

Instrumental Intimacy

EEG WEARABLES & NEUROSCIENTIFIC CONTROL

MELISSA M. LITTLEFIELD

Johns Hopkins University Press

BALTIMORE

Johns Hopkins University Press
2715 North Charles Street
Baltimore, Maryland 21218-4363
www.press.jhu.edu

Library of Congress Cataloging-in-Publication Data

Names: Littlefield, Melissa M., 1979–, author.
Title: Instrumental intimacy : EEG wearables and neuroscientific
  control / Melissa M. Littlefield.
Description: Baltimore : Johns Hopkins University Press, [2018] |
  Includes bibliographical references and index.
Identifiers: LCCN 2017020241| ISBN 9781421424651 (hardcover :
  alk. paper) | ISBN 9781421424668 (electronic) | ISBN 1421424657
  (hardcover : alk. paper) | ISBN 1421424665 (electronic)
Subjects: | MESH: Electroencephalography—instrumentation |
  Monitoring, Physiologic—instrumentation | Biosensing
  Techniques | Electroencephalography, Ambulatory
Classification: LCC RC683.5.E5 | NLM WG 26 | DDC
  616.1/207547—dc23
LC record available at https://lccn.loc.gov/2017020241

A catalog record for this book is available from the British Library.

*Special discounts are available for bulk purchases of this book. For more
information, please contact Special Sales at 410-516-6936 or specialsales@
press.jhu.edu.*

Johns Hopkins University Press uses environmentally friendly book
materials, including recycled text paper that is composed of at least
30 percent post-consumer waste, whenever possible.

To my dad, John Littlefield,
for encouraging me to do what I love and love what I do

The future is intimate. We want to be closer to ourselves.

—KRISTIN NEIDLINGER

Contents

Figures

Acknowledgments

Book writing can feel like a solitary affair, taking place, as it often does, at 5:00 a.m. or in the wee hours of the night while children sleep. Thankfully, when I did find myself in the company of others, I was encouraged and supported by many living, breathing human beings. I would like to first acknowledge and thank the members of my writing group, Jenell Johnson, Kelly Gates, and Anne Pollock, for their intelligent and critical reading, sharp wit, and good humor. I want to say a special thank-you to Jenell for listening to endless iterations of this project before it was ever born. For her unfailing support in all things academic, my thanks go to Rajani Sudan, *Configurations* coeditor extraordinaire. My appreciation is extended, as always, to Susan Squier for her mentorship, willingness to explore the horizons of academic thinking, and sage advice. Thanks to those whose work in all things related to the neurosciences continues to inspire me: Anne Beaulieu, Felicity Callard, Suparna Choudhury, Joe Dumit, Des Fitzgerald, Giovanni Frazzetto, Kelly Joyce, Laura Otis, Martyn Pickersgill, Andreas Roepstorff, Nikolas Rose, Jan Slaby, and Elizabeth Wilson.

I am fortunate to be surrounded by excellent colleagues in the field of literature and science at the University of Illinois, including Bob Markley, Lucinda Cole, Gillen Wood, and Bruce Michelson. For the two wonderful years that I led the Network for Neuro-Cultures INTERSECT Graduate Training Program, I was surrounded by colleagues and graduate students from across the university and immersed in excellent conversations about the neurosciences, experimental design, and transdisciplinarity. And I appreciate that Rex Ferguson, James Purdon, Simon Cole, and the members of the Art of Identification Network helped me to keep one foot in the forensic sciences while I was working on this EEG project. Thanks to the members of the Society for Literature, Science, and the Arts (SLSA); the Interdisciplinary Humanities Center at the

University of California, Santa Barbara; and the faculty and students at the University of Texas at Dallas who listened and responded to portions of the book. I am grateful for the work of several excellent graduate student researchers at the University of Illinois, whose work was made possible by the Department of English and the UIUC Research Board: Eman Ghanayem, Brandon Jones, Logan Middleton, and Anna Robb. And, of course, many thanks to the anonymous reviewers of this work.

To my editor at Johns Hopkins University Press, Matthew McAdam, thank you for believing in my vision for this project. Special thanks to Deborah Bors, Catherine Goldstead, Juliana McCarthy, and the wonderful editorial, marketing, production, and design staff at the Press—you have all been a joy to work with. Glenn Perkins did a remarkable job copyediting the manuscript. I would also like to express my gratitude to several EEG wearable companies and independent artists for allowing me to reprint their images in these pages. Thanks especially to Arlene Ducao, Kristin Neidlinger, and Christian Nold. The cover image for the book was generously provided by Josh McKible of MCKIBILLO.

Beyond the academy, I have benefited from the support and guidance of so many family and friends: Doris Perialas, Jonathan and Sheila Littlefield, Jean Trujillo, Jen Crabtree, Marie Schaffner, Phineas Reichert, Alex and Kris Perialas, Ruth and Sheryl Littlefield, Steve and Amy Cartwright. I am also thankful to have had an excellent friend and interlocutor in the late John Decker. My mom, Valerie Perialas, is always listening on the other end of the line when I need an ear and good advice. Mom, thanks for being my first feminist hero. To the members of my spinning and knitting crew, especially Debbie Mandel, Beth Engelbrecht-Wiggans, Cathe Capel, Sue Cutter, and Dottie Wolgemuth, thank you for all of the creative joy you've brought into my life. Thanks to my son, Isaac Sosnoff, for the love and the laughter—always. Zach, you grew into a fine young man during the writing of this book; I am and will always be so very proud of you. To my dear partner, Spencer Schaffner, thank you for making our life together into a never-ending adventure; you have taught me so much about living fully and with good humor. This book is dedicated to my father, John Littlefield (aka the King). Dad, you are generous, thoughtful, kind, and disciplined—all the things I strive to be. A single book dedication doesn't do you justice, but it's a start.

Instrumental Intimacy

A "Machine to Guide Them"

Theorizing Instrumental Intimacy through Mobile EEG

In an era replete with brain-imaging technologies, electroencephalography (EEG) is one among many instruments that embody the desire to visualize and control the brain, to capture its minutest signals, project its varying levels of arousal to scientific and lay audiences alike, and link its signals to potential behavioral change. However, and in contrast to the more visually alluring technologies of positron emission tomography (PET), functional magnetic resonance imaging (fMRI), and magnetoencephalography (MEG), EEG has been largely overlooked in critical studies of the neurosciences.[1] EEG may not be dominating scholarly conversations, but in the shadow of its brain-imaging brethren, it has been undergoing a quiet metamorphosis: leaving the laboratory and infiltrating the popular wearables market.

We might look, for example, to iBrain, a proprietary EEG headset that the NeuroVigil corporation began distributing to governmental agencies, pharmaceutical companies, and academic research laboratories in 2009. iBrain is a new kind of EEG device that can measure the electrical activity of the brain without gel, wires, or a laboratory setting. Indeed, it is a single-channel *mobile* EEG device, impregnated with hidden electrodes, which looks and feels like a simple headband. iBrain is intended to improve the ease of EEG data gathering, create new archives of neuro-data, and move the technology of EEG out of the laboratory and seamlessly into users' lives.

NeuroVigil's product "stands out," the company claims, because "we are using EEG as a microscope, by creating maps of brain activity rich in biomarkers from single-channel high-resolution data recorded from the comfort and privacy of people's homes" (NeuroVigil 2016). By analogy—and mixed metaphor—NeuroVigil compares EEG to both a microscope and a mapmaking device, neither of which captures the scope or technical specifications of EEG, but each of which reveals important assumptions about how EEG is expected to function in science and society. The reference to microscopes is reminiscent of earlier psychophysiological analogies for technologies that were expected to reveal the body's internal secrets.[2] The reference to mapmaking concerns the conquest of territories not yet claimed, or even identified, and it is also reminiscent of the language used to describe brain-imagining technologies (said to "map" the brain) and their collected data (often called "atlases").[3] By using both the microscope and the map as analogs for EEG, NeuroVigil reveals latent hopes for and expectations of the technology: that we could use it for visualization and conquest, display and control.

NeuroVigil is among a new group of companies, including NeuroSky and EPOC, selling headsets that bring EEG into new spaces and social configurations.[4] Other sensors have certainly been transformed at the hands of the wearables industry, growing smaller and more fashionable with each iteration, but EEG has undergone perhaps the biggest shift—from wet, technical laboratory experimentation to portable headgear, from gelled and wired caps to baseball hats. Consumers can now buy an array of EEG-based devices intended for meditation and mindfulness (Muse, Melon, Lotus, Puzzlebox Bloom); hats and visors for fatigue management (SmartCap); audio headsets and eye masks for improved sleep (Kokoon, Neuroon); bike helmets for urban neuro-mapping (MindRider); gaming and creation-centered wearables (Neuro Turntable, Mindwave); memory consolidation headsets (NeuroCam); and fashion accessories that display one's mental arousal (NEUROTiQ, Shippo, Necomimi). Based on this list, the market share for and desirability of EEG wearables has become a speculative reality. This potential boom is being driven, at least in part, by companies such as NeuroVigil who actively imagine scores of consumers navigating to their FAQ page to find answers to their burning questions: "Can I buy iBrain?" or "Can we get our hands on the technology for free to experiment with it?" Whether we will all be donning EEG headsets in the near future is yet to be seen, but the newfound abundance of EEG wearables is both product and producer of a market for mechanistic guidance.

Following in the footsteps of other psychophysiological technologies, EEG wearables are becoming the latest instruments through which we recognize our states of mental arousal and attempt to control ourselves more efficiently. Direct-to-consumer EEG wearables offer to identify, visualize, and render modifiable various aspects of our brains: they claim—as we see in the case studies that follow—to be able to help users sleep better, achieve more, get into and stay "in the zone," exercise brain "muscles," tag memory maps, and display precise levels of arousal for others and themselves to see. EEG wearables are legible within a consumer market defined by and obsessed with perpetual training, tracking, and competition, a market embedded with cultures of risk that demand even (and especially) the "healthiest" among us to subject ourselves to the oversight of various (medical) experts.[5]

The case studies in this book prompt two related arguments: First, historical, popular, and scientific narratives have constructed EEG as an ideal, mechanistic interlocutor for our brains. In other words, EEG is a product and producer of the belief that we need a machine to guide our behavioral modifications, calibrate our levels of arousal and activity, and mediate our interactions with the world around us. In its latest—wearable—incarnation, EEG promises to intervene in nearly every aspect of contemporary life, from sleep to athleticism, from fashion to travel. Second, EEG's early-twentieth-century emergence, mid-twentieth-century applications, and twenty-first-century transformations have brought into being the very states of mind that the technology is intended to monitor and modify. Before Hans Berger's experiments the 1920s and 1930s went public, physiological states were not discussed in terms of brain activation. The best examples are sleep and death, which had very different discourses and theories attached to them before we could record brain activation and use it as a clinical factor to determine those states. By analyzing EEG wearables in light of scientific and popular discourses from the twentieth century, I aim to shed light on the contemporary trend of self-tracking, the long-standing practice of using EEG to monitor and manipulate behavior, and several ideological beliefs about brain control that emerged a century ago and persist into the twenty-first century.

In these pages, I look backward in order to look forward, analyzing the burgeoning consumer market for mechanistic access while also challenging the novelty of brain optimization discourses emboldened by EEG wearables. This book is both a history of EEG as a technology for instrumental intimacy and a set of case studies that theorizes EEG's contemporary consumer market. Each chapter concentrates on a set of EEG wearables that claim to display, monitor, or intervene in particular brain states. By engaging both EEG's present and its past, my intention is to trace the emergent genealogies of various states of mind (from arousal to expertise to sleep, even to death), challenge the conflation of electrical activity with "brain waves," and analyze the marketing of mechanical brain monitoring. EEG wearables are not only devices for self-tracking; increasingly, they are coming to define who we are, how we know ourselves, and how we relate to one another.

Inventing Brain Waves

Even if you don't know much about human electroencephalography, aka EEG, most of us use and recognize the term *brain wave*: for example, "it was amazing to see my brain waves on the hospital monitor!" or "I'm trying to practice meditation so that I produce more alpha brain waves," or even the recent headline, "Scientists Found a Way to Email Brain Waves" (Pearson 2014).[6] In science fiction, popular science articles, and scientific papers, the term *brain wave* has become a colloquialism for telepathic transference (thoughts made material), for the spark of inspiration, and, since Berger's twentieth-century experiments, for the results of electroencephalography.[7]

When I first began working on this project, I had to remind myself that brain waves are an invention, one that capitalizes on a particular construction and conglomeration of electrical brain monitoring (primarily because the oscillograph recording of EEG looks like a pattern of waves) and our imaginative expectations—that is, that the brain is emanating its status and information in a material form.[8] Thus, if the advent of the EEG rendered visible our brains' electrical potentials, it also revealed the ways that "inventor" Hans Berger—and now many of us—conceive of the materiality of psychic processes. I discuss the rise of material thought elsewhere (Littlefield 2010, 2011); one of my primary assumptions is that we cannot take EEG for granted as a transparent technique that renders the brain objectively visible. EEGs do not measure "brain waves" any more than our brain emits them. Instead, EEG, including its newest, wearable form, is a product and producer of "brain wave" ideologies—what I define somewhat cheekily as the (electrical) *potential* of our brains to produce measureable states of mind and be open to intervention.

Understanding brain wave ideologies necessitates that we open the black box of EEG and consider its genealogy. The standard narrative describes EEG's emergence as follows: Human electroencephalography—the basic system on which all EEG wearables are based—emerged in the 1920s and 1930s as a means of collecting and recording the electrical activity of the human brain.[9] Obtained via the scalp, these measurements were some of the first noninvasive techniques that revealed when and how the human brain was an active organ.[10] Hans Berger, a German psychiatrist, was the first to record the human EEG;

he wrote of his trials in fourteen papers between 1929 and 1938.[11] His equipment varied, and he used instruments technically intended for cardiographic rather than cerebral measurements.[12] Nonetheless, Berger reported that "the electroencephalogram represents a continuous curve with continuous oscillations in which, as already emphasized repeatedly, one can distinguish larger first order waves with an average duration of 90 σ and smaller second order waves of an average duration of 35 σ" (Berger [1929] 1969, 70).[13] Berger named this first recorded rhythm, captured when the subject's eyes were closed, "alpha," and the second recorded rhythm, captured when the subject's eyes were open, "beta" (Ahmed and Cash 2013, 2).[14] Despite setting out to disprove him, E. D. Adrian and B. H. C. Matthews (1934) replicated Berger's results, offering validation to what they termed the "Berger rhythm."[15]

This version of the Great Man discovery narrative focuses only on a portion of a much more complex story of scientific creation, and it does so through a particular ideological lens. It does not, for example, call into question "discovery" as a construction, nor does it question what EEG was, is, or could become. Casting a broader net, as many historians of science do, we might note that EEG's conceptual genealogy also includes telepathic conjectures of brain-to-brain communication, experiments involving psychical energy, and the hope of creating a *Hirnspiegel*, or "brain mirror" (Gloor 1969; Millett 2001).[16] Moreover, Hans Berger's work on the EEG was hardly acknowledged in his native Germany for almost a decade, and he faced criticism abroad because the concept behind EEG itself—that the brain was an electrically active organ that could be accurately monitored—was quite radical.[17] As Peter Gloor notes in the introduction to his translation of Berger's initial papers on the electroencephalogram, "The publication in the 1920s of the first paper on the human electroencephalogram by Hans Berger was an event for which the scientific world was not prepared" (1969, 1). In his preface to Gloor's translations, Herbert Jasper retrospectively explains Gloor's claim, noting that "it seemed highly unlikely at that time that the simple rhythmic waves, the 'Alpha- and Beta-Wellen' of Hans Berger, could possibly represent the true electrical activity of such a complex nerve tissue as the cerebral cortex, especially in man, recorded not by an experienced electrophysiologist but by a psychiatrist with rather crude and simple apparatus" (v). Beyond being ostensibly out of his depth, Berger was fundamentally out of step with the science of the time, and the concept of accessing and mapping something as complex as the human central nervous system seemed outside of the purview of his work.[18] As Robert Kaplan argues, "So great were regarded the obstacles of tracing the electricity of the brain at the time that anyone trained in the field regarded it as futile" (2011, 169). In short, what Berger was attempting had a grounding only in animal experiments; it was believed to be utterly unattainable in an organism as complex as the human.

Despite the authority of dominant paradigms, Berger's experiments did change the shape of neuroscientific research, introducing a novel means of reading the electrical impulses of the brain for information about the mental state of the subject. Knowing about Berger's struggles for recognition and the possible alternative experimental and explanatory options that ended in EEG helps illuminate some of the imaginative potential embedded in the technology. Indeed, some of that potential still exists in EEG wearables: in the popular construction of EEG as "the brain machine"; in the colorful representations of arousal in EEG headsets (see chapter 1), and in the continued hopes for telepathic communication and telekinesis.[19] As Wolpaw and colleagues note in their work on brain-computer interfaces (BCIs), "People have also speculated that the EEG could have a fourth application, that it could be used to decipher thoughts, or intent, so that a person could communicate with others or control devices directly by means of brain activity, without using the normal channels of peripheral nerves and muscles" (Wolpaw et al. 2002, 768). Beyond speculative dreams, and at a very basic level, that we now speak of brain waves as if they are palpable, explicable, and accepted is attributable to Berger's work on oscillations, waves, curves, and the idea that the human brain might be producing measureable—and meaningful—electrical signals.

Visualizations of the brain's electrical activity are products and producers of brain wave ideologies. We have used Berger's insights—and the ideologies surrounding the emergence of EEG—to change how we understand the brain: its levels of arousal, (athletic) training potential, sleep cycles, and memory consolidation. Nevertheless, as Nikolas Rose and Joelle Abi-Rached note, it has not yet "proved possible to articulate clearly the way in which brain states give rise to mental states, and still less how they enable conscious experience: despite the proliferation of terms to describe the relationship, the explanatory gap remains" (2013, 226). We should, therefore, remain skeptical of our drive toward instrumental intimacy. As a first step, let us pause to consider what has become accepted knowledge about the electrical impulses of the human brain.

EEG in the Twenty-First Century: Mechanism and Regulation

Today, researchers would largely agree that EEG ostensibly measures the aggregate electrical potential of neurons via electrodes placed at various points on the scalp. A typical description of what EEG captures reads as follows: "An EEG signal is a result of a number of concurrent neuronal activities occurring in our brain. They include our current mental state (changes) and cognitive processes, various external inputs to our senses and various internal inputs and outputs to our internal organs" (Mihajlović et al. 2015, 8). Note that this particular description links neuronal activity with a variety of processes, including "current mental state." Such associations between the mental and the material are reminiscent of Berger's initial theorizations about EEG's potential and the brain wave ideologies that have since developed. Currently, there are five accepted electrical brain potentials (often referred to colloquially as brain wave frequencies) used in EEG experiments: gamma, beta, alpha, theta, and delta (with gamma being the most active and delta representing deep, healing sleep). The relative level of arousal is a key marker for many EEG wearables (more on that in chapters 1 and 4).

Although ambulatory EEG applications are proliferating, laboratory preparations, which typically produce the best signal quality over the most channels, require extensive preparation of the subject's head. Because EEG measures minute electrical activity in and around the brain, more electrodes equates to better signal quality. To collect the best signals, a subject's scalp may be lightly abraded under each electrode, and electrolyte gel is typically applied to adhere the electrodes and bridge the signal gap between scalp and electrode. Limitations of the technology necessitate that the testing environment is controlled for various noises that can interfere with an EEG signal, including heartbeats, movement, and even overhead power lines. In addition, EEG recordings are affected by the placement of the electrodes (where on the skull—and therefore over which brain region—they are located) and by the number of channels collected. Likewise, EEG aggregates data and does not represent the electrical potential of specific neurons (Mihajlović et al. 2015, 8).[20] This means that algorithmic claims made by several of the products examined in this book are not necessarily supported by the available, wearable EEG technology but instead by ideologies about visualization and control.

Despite its limitations, EEG remains one of the most portable brain measurement tools that we have in medicine and in the private sector. Unlike fMRI, which requires a massive magnet and extensive safety procedures; PET, which utilizes an injectable radioactive tracer; or MEG, which requires a magnetically shielded environment, EEG is a rather safe and portable technology for imag(in)ing brain activity. We do not often think of EEG as a brain-imaging technology, in part because it does not produce the same kinds of visualizations, pictures, images, and "lit up" brains of fMRI and PET that are popularized in the media. However, in its newly portable incarnation, the visualization of EEG data is granting the instrument new life, particularly in a culture obsessed with self-tracking, social reporting (on various Internet sites), and data visualization. (I return to this important shift in the visualization of EEG data in chapter 4).

Until recently, EEG was *only* accessible in laboratory situations. EEG wearables are a relatively new technological phenomenon, emerging in the early 2000s. There are several varieties on the market, but generally speaking, they allow users to monitor very circumscribed aspects of the brain's electrical potentials through a portable device worn on the head; these data can then be used for various purposes: they can be collated and displayed via a smartphone application, through which users can track their progress over time; they can be used to control on-screen games; or they can be used to control the (often visual) output of various devices (see chapter 1). Often, as is the case with most wearables, the device is fashionable and discrete: a headband or a streamlined headset that would draw little attention to the user.

EEG wearables, like most other wearable devices on the market, are unregulated by the Food and Drug Administration (FDA), state health agencies, or the Center for Devices and Radiological Health (CDRH) in the United States; abroad, wearables do not yet face any more scrutiny than they do in the United States.[21] The FDA, which became initially involved in the monitoring of medical devices in the 1970s after a spate of injuries and fatalities, have not yet brought most wearables under their regulatory umbrella. As of 2016, many wearables—including those that rely on EEG—continued to fall outside of the FDA's two primary triggers for regulation: being categorized as a "medical device" or creating "risk" for users (CDRH 2016, 2–5). In her examination of the FDA's 2016 draft document, "General Wellness: A Policy for Low Risk Devices," Anna Wexler notes that "the definition of a medical device is not based on the mechanism of action of the device, but rather on its intended use: a product is a medical device if it is intended for use in diagnosis or treatment, or intended to affect the structure or function of the body" (2015, 677). As long as their advertorial claims avoid discussions of treatment and disease and the product is not invasive, implanted, or risky, then it does not require regulation because it falls under the "General Wellness" category. "Many consumer EEG devices . . . seem to have strategically marketed their devices for 'improving mental fitness' and 'optimizing brain performance'" (682). The Melon headband, for example, is intended to be used for mental fitness training, meditation work, stress reduction, and productivity reports—with readouts of brain electro-activity managed through one's smartphone. Melon joins several other devices already on the market, including Mindo, and Muse.

Machines, Arousal, and Observation

Technologies such as EEG wearables are products and producers of *instrumental intimacy*, a means by which we learn about, access, and manipulate ourselves (in this case our brains) by interfacing with machines. Instrumental intimacy entails (re)imagining psychological or behavioral ontologies as physiological substrates; mechanically accessing, measuring, and comparing those physiological states to create recognizable states of mind; and creating paradigms for optimization. Consumer discourses surrounding EEG (and its wearables) encourage users to view the technology as helpful: it is not working against them to reveal some inner secret but *with* them to produce a more accurate portrait of their arousal and, by extension, their state of mind. At the same time, users are buffered from themselves by the technology: their levels of arousal are registered by the wearable, but knowledge and control are not always in the same hands. Wearers are both the observed and the observer, having access to *mediated* aspects of their own data (most of which has been preprocessed by an algorithm).

Arguably, instrumental intimacy depends on an earlier paradigm shift from "embodied-feeling" to "machine-mediated seeing" (Dror 2011, 329). This process took place in the laboratories of the early to mid-twentieth century, but it had ramifications for the ways that lay audiences understood their emotional capacities relative to the machines and technologies they encountered. In the laboratories, this shift from embodied experience to mechanistic records enabled scientists to work on dangerous subjects, such as emotion and arousal, by interjecting graphic inscription technologies between themselves and their subjects. In the field, adoptions of machine-mediated seeing meant that one's embodied experience was not always ratified by scientific experts and that experiencing certain, extreme emotions began to go out of fashion. As both Peter Stearns (1994) and Brenton Malin (2014) have argued, one affective ideal for Americans during the period between 1920 and 1960 was emotional "cool."

One of the key components of this shift from embodied feeling to machine-mediated seeing during the late nineteenth century and into the early twentieth century was the development of graphic inscription technologies designed to "read the body" or "read the mind" through the body. Very basically, graphic inscription technologies charted some aspect of one's body—typically an aspect of one's physiology—and most graphed the results using one or more styluses on various surfaces, from smoked drums to paper. Among the more famous are the plethysmograph, which measures volume (in a limb, for example); the sphygmanometer, which measures blood pressure; the pneumograph, which measures respiration; and the galvanometer, which measures electrical skin conductance.[22] When combined in various permutations, one might recognize these as components of some of the more infamous graphic inscription technologies, including the polygraph (sometimes colloquially called the lie detector). Quite often, these recordings of physiological processes were intended to measure a subject's physiological arousal, and then various levels of arousal were equated with emotional states. EEG emerged during this same era, traced its measurement of the brain's electrical potentials onto paper, and helped to equate states of arousal to various states of mind. The advent of EEG revealed electrical brain patterns that were literally and figuratively inaccessible to individuals without the help of a machine. Not only did EEG mediate the experience of the brain's electrical activity, but without or before EEG the brain's electrical activity was *fundamentally* inaccessible because it had not yet been imagined into existence.

Graphic inscription technologies were popular for at least two related reasons: they seemingly subverted a subject's conscious control, and they avoided the potential subjectivity of the investigator. On the first count, graphic inscription technologies measured aspects of human physiology that people supposedly were unable to consciously influence. This meant that scientists considered them to be more accurate, less subject to personal interpretation, and closer to nature. As Otniel Dror explains, experimenters using graphic inscription technologies to measure various emotions

argued that the viscera reflected a truer, deeper essence of emotion than methods that depended for their observations on the conscious awareness and self-reports of subjects or observers, or systems that depended on "superficial" facial expressions and gestures. They also suggested that observations of visceral emotions were easily converted into quantitative measurements, in comparison to the difficult process of quantifying verbal descriptions or facial expressions of emotions . . . that visceral emotions were independent of language and were hard to manipulate or skew deliberately; and that the involuntariness of the visceral emotions presented a "natural" distinguished from the artificial/mannered/social expressions and gestures of emotions. (2011, 327)

Dror details the myriad reasons investigators would prefer to put their faith in a machine as opposed to an introspective subject. But this faith in machines and the search for objectivity also extended to the investigators themselves.

In *Feeling Mediated*, Brenton Malin argues that investigators wanted to do away with their "burdens of intimacy" (2014, 183)—the taint of subjectivity that they began to catalog in studies of affect—or what Dror calls the "embodied-feeling paradigm" (2011, 329). As a solution, the "sociological research of the 1920s and 1930s saw a gradual movement towards 'scientism' and quantification as well. . . . Sociologists and psychologists aimed to demonstrate their objectivity by showing that they were beyond the emotionally tainted pasts of their own disciplines as well as above the emotional stimulations of the new media age" (Malin 2014, 164). In this scenario, machines provided a necessary solution: investigators could use them as an intermediary, a screen, an absorptive mediator that would leave them unscathed by subjectivity, emotion, or any intimacy with the subject. "Freeing themselves from emotional involvement, these researchers ascribed to their machinery the most empathic access to their subjects' intimate feelings. . . . By assigning to the apparatus the burdens of intimacy, scientists could frame themselves as distant bystanders appropriately detached from their subjects" (183). For psychophysiologists, Dror explains that one of the benefits of this machine transposition was the maintenance of a masculinized profession, as traces of the feminine were foisted onto receptive machines: "instead of becoming vessels for or embodying feelings (and 'reading minds'), they *observed* feelings through technologies, which embodied the traits of sensitivity, passivity, and impressionability, to do women's work for them" (2011, 341). This objective distance was but a ruse; nonetheless, it helped usher in a new mode of experimentation: what Dror calls the "machine-mediated seeing paradigm" (329).

Once this "burden of intimacy" is transferred to a machine, there is a paradigm shift in expectations about scientists'—and eventually our own lay—capacity to feel. One of the implications is a newly developed ideological incapacity to register physiological, psychological, and emotional changes for both researcher and subject. Only the machines have access: "The machines knew an individual's feelings better than the researcher, indeed, better than the feeling subject himself or herself, identifying emotional changes far too subtle for the subject to notice" (Malin 2014, 183). We see the extension of this assumption in lie detection paradigms (Littlefield 2011) and, later, in scientific experiments concerning the limits of "free will." In 1983, for example, the potential sensitivity of machine over introspective reporting was demonstrated anew by Benjamin Libet and colleagues. Their study "Time of Consciousness Intention to Act in Relation to Onset of Cerebral Activity (Readiness-Potential): The Unconscious Initiation of Freely Voluntary Act" is often cited by scientists and philosophers alike as a sticking point in debates about free will. If, as the study seems to demonstrate, "the brain evidently 'decides' to initiate, or, at the least, prepares to initiate the act at a time before there is any reportable subjective awareness that such a decision has taken place" (Libet et al. 1983, 640), then are we willing to extrapolate from the experiment's conditions to real-world scenarios? Or is the definition of free will too circumscribed for general use? Libet and colleagues argued that there were limits to the applications of their study, yet it remains central to ideologies of the instrumental intimacy in which we live in the twenty-first century. Acting in their newly minted observer role, machines "patiently recorded nature's answers verbatim: 'The observer no longer reasons; he registers'" (Daston and Lunbeck 2011, 4). And it is this role, as observer or registrant, that EEG wearables are expected to fill; users of these technologies are *volunteering* to have their electrical brain potentials recorded so that they might gain access to an organ with which they have been told they are out of touch.

In the twentieth century, the shift from embodied feeling to machine-mediated seeing granted knowledge—and therefore power—to the experts, meanwhile creating a new lay understating of any visceral connection to one's body: "What had been taken for granted, the natural ability to read emotions, was now represented as requiring expertise and technologies" (Dror 2011, 336). In the twenty-first century, EEG wearables retain a certain demand for expertise—we see this most clearly in the proprietary algorithms of the EEG wearables that are not available to the public user—but they have also transferred the "burden of intimacy" and option of instrumental intimacy to that user. Lay audiences for EEG wearables are being asked to think differently about (what have been defined for them as) their states of arousal. And this new form of experience, like the laboratory experiences of the last century, is demanding a new way of seeing.

One type of feeling that has become specifically mediated by EEG wearables is *arousal*, a state that both Dror and Malin parse—for Dror it is adrenaline, or the blush; for Malin it is excitation and the dangers of excess feeling. But be forewarned: this is not a book about affect or affect theory.[23] Rather, I analyze arousal as a state of mind, a social construct, and a useful aggregation. In chapter 1, arousal levels construct users' brains as attentive, tired, meditative, or relaxed; in chapter 2, activation (defined as changes in arousal) and theories of accelerated learning construct the brain as receptive for training; in chapter 3, constructions of arousal render sleep as an active state of mind that could be manipulated; and in chapter 4, the aggregation of arousal and activity patterns have the potential to create neurogeographies and consolidate memories. EEG wearables sell us a body that cannot regulate itself, one that requires machine-mediated seeing in order to measure, record, and modify our states of arousal and the states of mind that they diagnose; a body that is fundamentally defined by instrumental intimacy.

Brain Control: The Rise of Neuroscience and the Quantified Self

The instrumental intimacy afforded by EEG wearables depends on two assumptions: that we believe in our brains as a seat of control and a locus of identity and that we are willing to (and sometimes interested in) tracking our brains over time and as a means to effect change.[24] Not coincidentally, the rise of brain visualization technologies and the advent of self-tracking roughly coincide. fMRI was introduced in the 1990s, a decade that also saw the rise of self-tracking behaviors undertaken by individual artists, scientists, and lay people alike, behaviors that have since been solidified into the quantified self (QS) movement.

EEG wearables are, first and foremost, headgear. Whether they take the form of a headband, a baseball cap, a headdress, or headphones, the focus is on bringing the instrument into close contact with the scalp and underneath, the brain. While this makes a certain amount of sense—EEG sensors are, after all, expected to record electrical potentials from the brain—some of them fetishize the wearable aspect, drawing attention to their potential cyborg-like augmentation of the brain. Indeed, in their emphatic attachment to the head, EEG wearables imply that one does not simply control one's brain; instead, one must go through technological channels to gain access to the brain, and the potential to manipulate it. Moreover, EEG wearables offer ostensibly (more) minute ways of recording and manipulating the brain; corresponding discourses of control that explain this new access do so in terms of active management and intervention via technological mediation. Whether the brain is central to our conceptions of self, a fantasy presented by consumer marketers of neurotechnologies such as EEG wearables, or simply another organ in a complex physiological system, we are being asked—and are actively seeking out the means—to control our brains in ways that, at first blush, seem novel.

That we focus on the brain at all as the primary organ of control is both a long-standing belief and a specific product of our cultural moment. In his essay "Brainhood: Anthropological Figure of Modernity," Fernando Vidal argues that the neurosciences did not create brainhood (or brain-centrism) but instead took advantage of the already established belief that our brains are at the center of our selves: "A good number of 20th- and 21st-century neuroscientists seem to think that their convictions about the self are based on neuroscientific data. In fact, things happened the other way around: brainhood predated reliable neuroscientific discoveries, and constituted a motivating factor of the research that, in turn, legitimized it" (2009, 14). He goes on to argue that this belief has helped to craft the contemporary neurosciences "from 19th-century phrenologists palpating head bumps, through EEGs starting in the 1930s and up to today's brain scans, the hope of being able to read the mind and the self through brain recordings has not subsided" (19). This focus on the head—and brain—is evident in EEG wearables.

Vidal's exploration is one answer to Joe Dumit's excellent and haunting question about contemporary neuroscientific attempts to map the mind: "Why is it, that when we find a reading correspondence in the brain we are satisfied that we are in the right place?" (Dumit 2003, 39). On the one hand, brain-centrism helps to create a locus of action; on the other hand, it reduces psychology and conceptions of mind to a single organ, or, as Elizabeth Wilson notes, focusing on the brain "narrows the geography of mind from a diverse, interconnected system . . . to a landscape within which the brain, as sovereign, presides over psychological events" (2011, 280). EEG and its wearables not only take the centrality of the brain for granted, but they have helped to shift conceptions about the brain, rendering it an organ of interest. One of the most striking examples (and one I discuss in the conclusion), is the advent of brain death, which moved the locus of life from heart to brain.

Because of its constructed centrality, the brain is defined (and consumed) as both problem and solution. In neuroscientific studies, print news media, pharmaceutical commercials, and EEG wearable advertisements, our brains are often represented as troubled, out of control, untrained, and in need of technological intervention. The irony, of course, is that intervention often begins in and depends on that already problematic organ. In her study of brain technology users, for example, Jonna Brenninkmeijer argues that "technologies like EEG devices make clear that people are troubled by their brains, and that is why the self has to work on the brain to improve itself" (2010, 113). The idea of "working on the brain" assumes both access—granted by EEG's amplification of the minutest electrical signals—and malleability. Indeed, our hopes for brain control are mutually imbricated with theories of brain plasticity.

From the EEG-based bio- and neurofeedback studies that emerged in the 1950s to the colloquial expression "nerves that fire together wire together," the brain has been reconceived as a plastic organ capable of reestablishing connections when tissue is lost or rerouting its so-called circuitry when something goes wrong. Theories of plasticity have shifted locationalist conceptions of the brain into functionalist ideas, leading us to believe that we better understand some of the brain's robustness. But plasticity is also related to ideals about intervention, manipulation, and, ultimately, control. If the brain is malleable, and is a potential seat of the self, then changing one's brain becomes an acceptable route to behavioral, psychological, and physiological manipulation. According to Nikolas Rose and Joelle Abi-Rached, "brains are now constructed as persistently plastic, molded across the life-course, and hence requiring constant care and prudence" (2013, 223). This turn to "constant care and prudence" necessitates certain forms of (self) tracking, modulation, and modification.

The mainstream uptake of plasticity discourse has led, at least in part, to a surge of brain training, neuro-gymnastics, and other efforts focusing on brain health and optimization; these are linked in large part to specific technological interventions. As Simon Williams and colleagues have argued, "The growing industry of efforts to boost, improve or enhance cognition in various ways takes us beyond a concern with 'therapy' to the promise, prospect or temptation of mental optimisation that extend conventional definitions of health and illness, normal and abnormal, and function and dysfunction" (Williams, Higgs and Kats 2012, 74). Many of these efforts are aimed at the brain because of a powerful cultural belief that "the neurosciences hold out the promise of treating, managing, protecting, even boosting or improving the cognitive powers of the brain and thereby 'improving,' if not 'optimising,' the human condition in various ways" (66). Brain training, which is associated largely with pen and pencil games or computer software, is often targeted specifically at elderly users or users who fear the possibility of dementia in old age.[25] While EEG wearables are quite different from many brain-training programs (see chapter 2), ideologies of instrumental intimacy rely on similar logics of optimization.

In the popular imagination, plasticity has been taken up as a theme in (science fiction) novels and film. Consider, for example, Kyle Kirkland's 2013 novel *Mind Plague*, which explores the burgeoning NeuroTech industry and the rise of a related disease, Synapse Interruption Syndrome. While investigating the disease's epidemiology, neuroscientist Bailey Breege breaks into a colleague's office to listen to his audio diary. Through his recorded thoughts, Stephan Blanadi, a minor character in the novel, plays a major role in helping Breege determine how to thwart the epidemic. According to Stephan, the NeuroTech industry should "consider the placebo effect. Consider the gifted practitioners of meditation, who routinely achieve mental control over almost any physiological process. And consider how easy it is for everybody to have at least some measure of control over these processes when they receive feedback on how they're doing. Anyone can learn to control various physiological parameters with the aid of biofeedback—all they need is someone, or *some machine to guide them*, to tell them when they're getting better at it. People have even learned to control the firing of a single neuron! The brain can control itself. How marvelous!" (Kirkland 2013, 80; emphasis added). Breege, who does not find much of interest in Stephan's recorded thoughts (and notes only in passing a copy of William James's *Principles of Psychology* on his desk), goes on, nonetheless, to fight SIS by urging sufferers and caretakers alike to harness the power of their minds. Theories of plasticity meet ideologies of mechanistic control, leading to an excellent example of instrumental intimacy. According to the resident neuroscientists, "all they need is someone, or *some machine to guide them*," and therein lies the crux of bio- and neurofeedback.

Narratives about plasticity have become closely linked to discourses of control and responsibility. In her book, *Brain Culture: Neuroscience and the Popular Media* (2011), Davi Johnson Thornton expertly details the move from discipline (via Michel Foucault) to control (via Gilles Deleuze) in popular neuroscience and its relations to discourses of "health" and "optimization." And in *Neuro: The New Brain Sciences and the Management of the Mind*, Rose and Abi-Rached argue that "we are now acquiring . . . the obligation to take care of our brain—

and of the brains of our families and children—for the good of each and all" (2013, 223). Rose and Abi-Rached's discussion of obligation has a certain resonance with NeuroVigil's efforts to market the iBrain. NeuroVigil specifically links its EEG device to the monitoring of health and risk, noting that "two billion people worldwide are affected by a brain or nervous system disorder, including 100 million in America. NeuroVigil's non-invasive, wireless window into the brain maps brain activity and can offer insight into the diagnosis and treatment of conditions such as Alzheimer's, autism, cancer, epilepsy, depression, and schizophrenia." Control of the brain, therefore, is not simply a nice idea, but an "ethical *obligation*": "The ethical implications [of the iBrain] are instead ethical *obligations* to drive this technology forward with non-invasive medical applications in an effort to accomplish things like a robust communication so-lution for locked-in syndrome, the mobilization of paralyzed or otherwise im-mobile people, the demonstration of complex mental states previously unknown in non-human animals, and the early detection of debilitating diseases where preparation is key to treatment and manageability. As pioneers, leaders, and experts, we are always pushing this technological innovation towards the better-ment of humanity" (NeuroVigil 2016, emphasis added). NeuroVigil provides not only exigency—but also imperative—for the development of wearable EEG; moreover, the statement moves seamlessly from complex disease mitigation to preventative medicine and from recognized conditions to as yet unknown states of mind. The drive to produce a "demonstration of complex mental states pre-viously unknown" (in nonhuman or human animals) all the while working toward "the betterment of humanity" is the basis of innumerable projects in the history, and contemporary explosion of, EEG.

If neuroscientific technologies have ostensibly enlarged our access to the brain and shifted our definitions of responsibility, malleability, and control, they have done so during roughly the same decades in which many of us began tracking ourselves. From mealtimes to sleep times, in the form of photos, time stamps, and steps taken, tracking ourselves has become a normal activity (Berson 2014; Dow Schüll 2016a, 2016b; Lupton 2016; Neff and Nafus 2016). Anthropologist Natasha Dow Schüll argues that "in 1990, just as digital infor-mation and communication technologies were coming into widespread use, the French philosopher Gilles Deleuze suggested that the architectural enclosures, institutional arrangements, and postural rules of disciplinary societies were giving way to the networked technologies of 'control societies,' involving contin-uous coding, assessment, and modulation. The latter scenario bears an uncanny resemblance to the tracking intensive world of today, in which the bodies, movements, and choices of citizens and consumers are ever more seamlessly monitored and mined by governments and corporations" (2016b, 24). The quantified self movement officially began in 2009 when nearly two decades of individual practice was consolidated into a consortium by Gary Wolf and Kevin Kelly. We know from the work of artists such as Stephen Cartwright that self-tracking, on which QS is based, began much earlier.[26] And bio-mapping, which is often based on a form of self- or aggregate-tracking, has a long and varied history that became technologically based in the 1990s and 2000s (see chapter 4).

As a practice, self-tracking brings us full circle to some of the questions and issues concerning both graphic inscription technologies and neuroscientific conceptions of control. Take, for example, Gary Wolf's explanations of the quantified self: "When we quantify ourselves, there isn't the imperative to see through our daily existence into a truth buried at a deeper level. Instead, the self of our most trivial thoughts and actions, the self that, *without technical help*, we might barely notice or recall, is understood as the self we ought to get to know" (Dow Schüll 2016b, 25, emphasis added). Note his caveat that "without technical help" we would miss aspects of ourselves that we might at first dismiss as trivial but later see as important. Here, the self is being reconfigured in a largely physiological way, away from psychological depth; in his comments we might hear echoes of Otniel Dror's and Brenton Malin's arguments concerning "burdens of intimacy" and a move from embodied feeling to machine-mediated seeing. Artist Eric Boyd argues that self-tracking is "about introspection, reflection, seeing patterns, and arriving at realizations about who you are and how you might change" (25). Dow Schüll interpolates Boyd's comments by noting that "this intimate journey commences not with a turn inward but with a turn outward to the streaming data of a device: an extraction of information, a quantification, a visualization. Selftracking, following Boyd, renders 'an exoself, or a digital mirror; it lets you look at things you otherwise couldn't see using just your own eyes, and see yourself more honestly'" (25). In each example, technological mediation, one aspect of instrumental intimacy, is viewed as an integral part of the quantified self. "Without technological help," Wolf argues that he could not notice the minute movements of his seemingly trivial thoughts and actions; for Eric Boyd, technology provides "an exoself, or digital mirror" through which he can "look at things you otherwise couldn't see."

In these statements, we hear echoes of not only graphic inscription technology but also NeuroVigil's marketing copy: technologies that allow for self-tracking provide a metaphorical microscope that allows one to get in touch with things that would, to the naked eye, seem untraceable, untrackable, and unmalleable. But note, too, the move back toward introspection—an activity from which graphic inscription technologies shielded psychophysiologists. However, in this new instantiation, introspection is rewritten as *necessarily*, and always already mediated by the machine.

In order to understand EEG wearables, we must engage several debates in sociological and anthropological discussions of the neurosciences and the quantified self if we are to parse ideologies of brain control. Kyle Kirkland's philosophically inclined neuroscientist argued that brain control is accessible to anyone: "All they need is someone, or *some machine to guide them*, to tell them when they're getting better at it." EEG wearables are one manifestation of such machines, and by examining their genealogies, contemporary instantiations, and the speculative—and not so speculative—futures that they propose, we can better assess the instrumental intimacies we often take for granted.

Structure and Methodology

This book is organized into four chapters and a conclusion: Chapters 1 and 4 frame the book in terms of public displays of individual and aggregate arousal and their implications for visualizing, monitoring, and controlling bodies and their interactions with environments. Chapters 2 and 3 delve more deeply into genealogies of controlling the brain when and where new states of mind are introduced and made visible and accessible via mechanical intervention. Arguably, the identification of these states is intimately linked to our belief in the measurement and import of levels of arousal. The middle chapters detail aspects of EEG wearables that are mediated by experts, while chapters 1 and 4 focus more on users, inventors, and their interpretations of the ideologies explored in chapters 2 and 3.[27] Throughout each chapter, the concept of instrumental intimacy structures the discussion: How are wearable EEG technologies both product and producer of novel conceptions of self—including the types of information we can share with ourselves and others? How do these technologies create new responsibilities of control, efficiency, achievement, and self-knowledge?

Methodologically, I am a literature and science scholar who follows concepts and technologies around, into, and through scientific studies and as they emerge in the public consciousness. In this book, fiction (in the form of novels, short stories, plays or poems) is not the primary source material; rather, I treat print news media, scientific studies, product advertisement campaigns, and corporate briefs *as* literature to be read, interpreted, and interpolated as constellations of discourses. Each chapter includes at least two EEG wearables either proposed for market (prototypes) or available commercially. In many cases, the EEG wearables I discuss are speculative fictions that do not yet fully exist in our world but nevertheless give every indication of their material reality and material impacts on our conceptions of the brain, theories of arousal and control, and potential for increased instrumental intimacy.

In chapter 1, "Public Displays of Arousal: EEG Wearables and the Aesthetics of Transparency," I answer two questions: When and how did it become fashionable to publicly display the electrical activity of our brains? And what do such public displays reveal about the ideal twenty-first century brain? Borrowing Rachel Hall's "aesthetics of transparency," through which she theorizes the ideal, post-9/11 traveler, I argue that EEG wearables capitalize on and offer up an ideal twenty-first-century brain that values discourses of transparency. Just as other fashionable wearables have opened the body to visualizations of its physiological data, EEG wearables help to fetishize a brain that is open, accessible, visualizable, and modifiable. To make my case, I turn to three EEG wearables that represent a wide spectrum of availability, user-centeredness, and surveillance capabilities: NEUROTiQ, a high-fashion headdress that displays EEG data from an EPOC headset as glowing colored globules—much like a mood ring once did for body temperature; Necomimi, a pair of animal-themed ears that twitch based on one's EEG data; and SmartCap, an EEG-sensor-laden baseball cap used by mining companies around the globe to measure driver fatigue. In each case, the transparency offered by EEG wearables is always already a translation—one that we should carefully examine for clues about the ideal twenty-first century brain.

In chapter 2, "In the Zone: Better Training through Alpha and Beta Waves," I take EEG wearables back to the laboratory to examine how neurofeedback is being utilized for novel learning paradigms. Specifically, I address the work of three institutions: the Biocybernaut Institute, the Mindroom, and the Advanced Brain Training Laboratory. Each has received some popular press coverage, and each has a unique research agenda based on neurofeedback and accelerated learning. Turning to scholarship from Jonna Brenninkmeijer, Nikolas Rose and Joelle Abi-Rached, and Brad Millington and drawing from an archives of print news media, popular nonfiction, and laboratory studies, I theorize a major shift in our construction of the educable brain. From unconscious to conscious and controllable, our brain has been rewritten as the space for efficient training and the center of possibility for paradigms of accelerated learning. Wearable EEG capitalizes on both of these ideologies, in and out of the laboratory. After working through the intra-laboratory examples, I turn briefly to extra-laboratory applications: EEG wearables such as Lotus and Puzzlebox Bloom.

Chapter 3, " 'Such a Natural Thing': EEG Sleep Science in the Laboratory and the Bedroom," examines a state in which most of us spend about one-third of our lives. Once considered to be an inactive state of mind, sleep was reconfigured and recategorized via EEG in the 1950s and 1960s as an active state of mind, and one in which scientists (as well as laypeople) could intervene. Nathaniel Kleitman and Eugene Aserinsky, a University of Chicago scientist and his student, effectively changed the landscape of sleep research by recording what we now call REM sleep. The discourses of the 1950s and 1960s, in both scientific literature and in the print news media, continue to inform contemporary mobile EEG technologies aimed at helping users sleep better. Although we still do not know why we sleep or what is the purpose of the REM stage, our sleep patterns have become something we can ostensibly control and modify—even to the point of addressing jet lag and shift work sleep disorders. Contemporary EEG-based wearables such as Kokoon and Neuroon draw from the ideologies born of the decades between 1950 and 1970 and play on discourses of risk, health, and efficiency that came into being with REM sleep.

Where chapter 1 details individualized public displays of arousal, chapter 4, "Neurogeography and the City: EEG's Collaborative Cartographies," deals in the aggregate to address new technological conglomerations that help map EEG activation onto cityscapes, landscapes, and photographic records. EEG wearables have the potential to reprise our understanding of brain mapping, and the political, economic and social uses to which aggregate EEG data can be put. To make my case, I situate what I call "neurogeography" within the context of another, older practice, psychogeography and the Situationists' *dérive*, examining EEG wearables such as MindRider and Neurocam alongside the work of broader bio-artist collectives. I then analyze some of the specific texts and data that have emerged from MindRider and Neurocam to assess EEG-based emotional (arousal) cartographies of the city and to consider memory consolidation practices afforded by EEG wearables.

Finally, in the conclusion, "From Soufflé to Signs of Death: Instrumental Intimacy about Us, without Us," I return to my initial questions about the concomitant rise of EEG and our desire for a mechanical guide and intermediary, reprise my theory of instrumental intimacy, and look to some possible future directions for EEG wearables research, including consumer testing of likes and dislikes, and corporate employee testing. I also examine the historic context for the advent of brain death as a state of mind in which arousal is absent and impossible. In each example, EEG's ability to provide instrumental intimacy allows third parties to make decisions about users without their involvement, further complicating EEG's relationship to our personhood and neuroscientific control.

The case studies in this book present EEG's genealogical past and its speculative futures. Rather than buying into their advancement as an "ethical obligation," as NeuroVigil labeled it, we can use these technologies and their narratives to situate EEG wearables as complex cultural objects that raise more questions than they answer. How has EEG been transformed from a laboratory technology into a household product? Why has EEG been discursively figured as the best mechanical guide for interacting with our brains? What are some of the social, medical, and instrumental outcomes of introducing EEG to extra-laboratory contexts? Yet, I, too, can only speculate about where the ideologies informing EEG wearables will take us in the near and distant future.

On a final, and personal, note, over the course of preparing this book, reading and writing about various states of mind led to some introspectively startling moments. Despite my skepticism about EEG wearables and—more centrally, my constructivist bent toward questioning assumptions, definitions, and discourses—I found myself participating, sometimes less than consciously, in neurofeedback experiments of various kinds. More often than not, when I found myself staring at a blank page, I would turn on some "alpha wave" study music to get myself "in the zone." When writing the sleep chapter, I found myself adjusting my assumptions about sleep—and practices, too: testing various bedtimes, thinking (too much?) about whether I was producing ideal brain waves as I lost consciousness, and wondering how my delta rhythms and REM sleep patterns compared to the average. I imagined a world of SmartCap wearers as I taught my fifth apocalyptic science fiction novel of the semester, and with each walk I took in the evening, I wondered what it would be like to track my patterns, directions, and interests with a smartphone Neuro Tagging Map. Needless to say, increased self-consciousness has been a hazard of working on this project. Readers may find themselves tumbling down the same rabbit holes.

1

Public Displays of Arousal

EEG Wearables and the Aesthetics of Transparency

> When someone asks "How are you?" you can look at your ring
> and say "I'm feeling a little yellow this morning" or perhaps "I'm
> marvelously blue, thank you."
> —"CHANGES COLORS WITH YOUR MOODS"
> ADVERTISEMENT, 1970

In the 1970s, mood rings were all the rage. The novelty fashion accessory promised to reveal aspects of yourself to you, as one Australian advertisement explained, "You can now be more aware of your inner feelings, thoughts and emotions and how they are really affecting you. You can learn more about yourself" ("Changes Colors" 1970). The rings, which contain liquid crystals that change color with the wearer's body temperature, not only base the measurement of "inner feelings" on an aspect of physiology but also assume that wearers have a particular range of temperatures—and corresponding inner feelings. The Australian model I mentioned relies on a seven-color scale that ranges from onyx black to sapphire blue. (You're aiming to be not simply Bright Lapis Blue but Sapphire Blue: "The Ultimate, passionate, deeply satisfied.") The language used to describe these feelings is reminiscent of the era and the culture in which the rings were popular; in early decades of the twenty-first century, few of us would likely consider "good vibrations" to be the description of "ultimate" feeling.

Mood rings may have fallen out of fashion, but they are part of a genealogy of wearables that foster instrumental intimacy: knowledge about physical, mental, and emotional states that are rendered available to the user only through technological mediation. We now have available all manner of EEG wearables that claim to register the electrical activity of the brain and translate these data into information about arousal levels. We might imagine them as Fitbits for the brain, or as mood rings for the head.

Unlike the other EEG wearables that I examine later in the book, the examples in this chapter are not intended for training purposes or for aggregating data; instead, they might be best understood through what Rachel Hall has termed an "aesthetics of transparency" (2007, 321). Although Hall uses the phrase in the context of US homeland security measures, aspects of her theory are equally applicable to an American market fueled by self-tracking and obsessed with new definitions of mental fitness. An aesthetics of transparency is intended "to force a correspondence between interiority and exteriority on the objects of the preventative gaze or, better yet, to flatten the object of surveillance, thereby doing away with the problem of correspondence altogether." Those whose bodies are always already open for inspection can be characterized as being "transparency chic" (Hall 2015b, 128). In the case of EEG wearables, individuals are asked to broadcast what I call "public displays of arousal": the outward expression of inner physiological states that have been translated to provide meaning for the wearers and their watchers.

EEG wearables that promote an aesthetics of transparency illuminate what Brenton Malin calls "systems of emotional disposition" (2014, 248), and which I would specify as systems of *mental* disposition: the fashionable discourses about desirable states of mind that surround and inform particular types of communication technologies. In the case of EEG wearables, this means paying attention both to the brain states that technologies are purported to evoke and to the cultural discourses that champion and explain those brain states. By encouraging individuals to wear their brain states on their proverbial sleeves, EEG wearables prompt several questions: Why are we invested in displaying our states of mind—and doing so fashionably? What assumptions about neuroscience, arousal, and bodies inform the technologies and the translations of data that they display? How are these technologies (among others) modifying our understanding of what counts as an acceptable public display of arousal? I argue that EEG wearables sell us a body in which states of arousal have become the most diagnostic—and socially relevant—states of mind. As we calibrate ourselves, we are being recalibrated to embrace particular instantiations of human sensation and physiological response.

In this chapter we look at three EEG wearables that illustrate the spectrum of available uses from speculative bio-art technology to novelty party wear to disciplinary instrument: Sensoree's NEUROTiQ, a headdress that changes color to display different brain wave activity; NeuroSky's Necomimi, a pair of novelty animal ears that move according to changes in brain waves; and CRC-Mining's SmartCap, a baseball hat embedded with electrodes to measure mental fatigue. The design of each device and the designs of its parent company reveal that transparency is always already mediated. The discursive, interpretative moves on which transparency is built can illuminate our larger expectations for the ideal twenty-first-century brain.

Fashioning Transparency through Systems of Mental Disposition

Over the years in which EEG has become increasingly portable, we have also witnessed an explosion in wearables—clothing and accessories containing electronic devices that monitor various physiological functions while also looking fashionable. This trend promises to continue.

> According to a recent report by Gartner, a technology research and advisory firm, shipments of health and fitness tracking wearables are forecast to reach 91.3 million by the end of 2016. But while 19 million of these devices will be "smart wristbands" and 24 million will be "sports watches," the largest and fastest growing category is expected to be "smart garments," much like the tech-infused tennis shirt (which monitors heart rate, breathing and stress levels) that Ralph Lauren debuted on ball boys at this year's US Open. In fact, over the next two years, as shipments of sport watches and smart wristbands are cannibalized by the rise of multi-functional smartwatches like the Apple Watch, shipments of health and fitness tracking "smart garments" are forecast to explode from 0.1 million units in 2014 to 26 million units in 2016. (Kansara 2014)

The spate of novel wearables has prompted scholarly attention from anthropologists and sociologists alike, including Natasha Dow Schüll (2016a, 2016b) and Deborah Lupton (2016). With an eye to the longitudinal, Dow Schüll notes that wearables are not new but a reiteration with a difference: "While people have long used simple, analog devices to record, reflect upon, and regulate their bodily processes, use of time, moods, and even moral states (here we can list mirrors, diaries, scales, wristwatches, thermometers, or the lowly 'mood ring'), the past five years have seen a dramatic efflorescence in the use of digital technology for self-tracking" (2016b, 24).

This boom in wearables as mass-market commodities should also be situated in the longer history of fashion and technology integration. Designer Amanda Parkes reminds us that "fashion has a deep history with technology. Look to Hussein Chalayan, not Apple, for inspiration on the future of wearables. Look back to Bauhaus costumes and theatre. Or back to 1919 when [Italian poet, editor, and futurist Filippo Tommaso] Marinetti wore a lightbulb tie which flashed for emphasis during speeches, perhaps the original wearable" (Kansara 2014). We might remember that fashion has long been one of the technologies of the body. We developed various skins and cloths, along with various technologically inflected methods to augment, shape, and wear them. "Fashion is an extension of both the physical and the aesthetic body—and is one of the first ways that technology began to enter the physical body's 'space'" (Fortunati, Katz, and Riccini 2003, 5). The body's physical space has indeed been radically modified over the course of the past century as we have seen the emergence of the cyborg and the post-human—conglomerates of technology and biology, fashion and augmentation.

Take, for example, the humble, wearable wristwatch—which began to take on the shape we now recognize in the late nineteenth century (Freake 1995, 71). Although it was not tied to a smartphone and did not collect data from the individual user, it did bind the wearer to a larger system that measured, valued, meted out, and demanded punctuality; it was both the product and producer of a phase shift in the experience of time passing. Like contemporary wearables the wristwatch ostensibly provided more control to the wearer who was, in turn, inculcated into a larger system of discipline: "The wristwatch seemed to give the worker, more particularly the office worker, more control over his or her time. Still, this seeming democratization of access to the time also instituted what David Landes . . . calls time discipline, as opposed to time obedience . . . 'punctuality [that] comes from within, not from without'" (Freake 1995, 72). The internalization of time, how it's read, what it means, and how it affects one's interactions with others is just one of many examples from the history of technology, fashion, and the body. At each instance of added measurability, the body becomes more fragmented, traceable, and open to evaluations and newly assigned meanings.

Whenever we examine the work of technologies in a culture, Malin reminds us, "we need to be equally concerned with the systems of emotional dispositions that surround various communication technologies. What sorts of emotions are believed to be communicated by our various technologies, and what do these beliefs tell us about our assumptions about our own and other's feelings?" (2014, 248). In short, the acceptable technologies, instruments, and experiments of the era reveal more about the cultures in which they are embedded than about their own inherent qualities. We would be wise to remember, for example, that when it comes to EEG, there is not a consensus about how to gather, interpret, use, or store EEG data: "There are no large publicly available databases that contain EEG activity of the normal healthy population. Neither is there a consensus on how to store the EEG information and the context in which it is recorded, nor on what features to extract from EEG signals for characterizing mental traits and states of a person" (Mihajlović et al. 2015, 8). Without consensus concerning the data, it becomes even more important to consider the systems of *mental* dispositions that inform EEG wearables.

The current market places a premium on technologies that can mark and make publicly visible our states of mental arousal (often defined as attention, focus, or fatigue). These technologies aid in the creation of bodies attuned to various levels of stress and able to recognize when they need recalibration (see chapter 2), but these technologies also draw our attention to systems of mental dispositions: the ability and desire to interface with a machine so as to limit the burdens of intimacy and self-reflection or introspection, the acceptance of a social climate in which privacy is not a primary concern, and the move toward cultures of self-quantification and transparency in which mental machinations are fashion accessories and one's state of mind can earn one differential status in occupational and social circles.

The latest EEG wearables also place users into disciplinary regimes, this time in terms of transparency—a move that calls into question divisions between interior and exterior. We could argue that "the aesthetics of transparency establishes a binary opposition between interiority and exteriority and privileges the external or visible surface" over other, complex sociocultural indices of identity (Hall 2007, 321), but I prefer Rachel Hall's revised definition of the aesthetics of transparency, which is to "force a correspondence between interiority and exteriority" or "to flatten the object of surveillance, thereby doing away with the problem of correspondence altogether" (2015b, 128). One reason I prefer this updated definition is because EEG wearables are Latourian black boxes through which data measurement, analysis, and output appear seamlessly related so that exploring the complex problem of correspondence between input and output appears to be unnecessary. There cannot be a problem of correspondence when the output is the only and natural result of the input; the exterior *is* the interior.

I also prefer Hall's revised definition of the aesthetics of transparency because of the interesting resonances it creates between theory and object. Consider one of the latest EEG wearables. Kristin Neidlinger, who created NEUROTiQ, explicitly characterizes her fashion wearables in terms of "extimacy," a term she derived from "externalized intimacy" and a concept she sees as emerging from the recent spate of social media and also responding to the needs of individuals with sensory processing disorders (SPDs). She does not link it to Jacques Lacan, to whom the term is generally attributed and who used it as a means to unsettle the divide between internal and external, intimacy and externality: "extimacy indicates the nondistinction and essential identity between the dual terms of the outside and the deepest inside, the exterior and the most interior of the psyche, the outer world and the inner world of the subject, culture and the core of personality, the social and the mental, surface and depth, behavior and thoughts or feelings. All expressions of the duality exteriority-intimacy would be hypothetically replaceable by the notion of extimacy, which precisely joins ex-teriority with in-timacy, and states explicitly the interpenetration and mutual transformation of both spheres" (Pavón-Cuéllar 2014, 661). Lacan's term is an apt one for wearables in general, and for EEG-based wearables in particular. Extimacy assumes that any break between the inside and outside, the internal and external is simply an illusion—a false binary that creates idea(l) extensions such as privacy and the public sphere. EEG wearables draw our attention to the liminal space between inside and outside, rendering the transition point—be it our scalp or our brain or our mind—as a public inter-face. In some respects, EEG wearables simply give us a device to stand in for or represent the always already breached border of consciousness. Likewise, Hall's reprised definition highlights that transparency is based on this forced correspondence between inside and outside.

From the seemingly direct correspondence of NEUROTiQ to the strange animal-like data translations of Necomimi to the numeric systems of SmartCap, EEG wearables are products and producers of Hall's "transparent chic." In Hall's arguments, the term implies specific class distinctions for individuals moving within systems of homeland security; in the case of EEG wearables, the labor of transparency enables different kinds of social distinctions. Each technology expects users to produce and allows them the "opportunity for performing" (Hall 2015, 25) desirable mental states demanded in particular social and occupational spaces.

Moreover, discourses of transparency as a technologically mediated ideal are related to Hans Berger's own desires for EEG. Perhaps because it represents unseen electrical potantials, EEG has embodied, and continues to embody, some speculative ideals for direct communication between brains (or between brains and machines). When Hans Berger first conceptualized EEG in the 1920s, he imagined it as a *Hirnspiegel*, or "brain mirror"—as a means to access the minds of the subjects in his laboratory (Millett 2001, 535) via a one-to-one correspondence between the psychic and the material (see introduction). He also imagined EEG as a type of telepathy, an ability to share one's mind with another.[1] In some respects, the recent spate of EEG wearables is the realization of Berger's brain mirror dreams: as one reporter notes of NEUROTiQ, "while it's not at the level of telepathic communication by any means, it is certainly a new concept in fashion design" (Krassenstein 2014).

Repackaged for the Runway; or, How EEG Went Public

The potential marketability and popularity of EEG wearables depends, in part, on an affiliation between fashion and technology—between the portability of EEG, the marketing of wearables for a particular demographic, and their desirability to a mass market. In an essay for *Fast Company*, Sindya Bhanoo (2014) discusses several advances in medical wearables, from tattoos to pills, noting that the key to selling wearables is not the novelty of the technology but the novelty of its application and (re)packaging. "The technology itself is nothing new. It might include, for instance, an accelerometer, a gyroscope, and an electromyography sensor to record muscle firings. This is about repackaging existing technology in a way that encourages regular, consistent use." Likewise, EEG is not a new technology, but its most recent applications manifest the power of repackaging.

NEUROTiQ, Necomimi, and SmartCap come from three distinct areas. NEUROTiQ represents high fashion: a prototype of the headdress graced runways in 2013–2014, but no model is available for sale.[2] Necomimi is fetishistic, costume wear, a novelty item available for general purchase but likely to remain in niche markets. The SmartCap is a mainstream accessory: a baseball cap that is widely used in at least one market with promise of global distribution. What draws these three together is an explicit attention to the fashionability of the product, its appeal to a particular demographic, and the explicit (re)packaging of EEG for a new consumer market or audience. This range of products—from esoteric to practical—helps illustrate the potential or speculative extrapolation of fashionable, EEG wearables: their widespread distribution, their affiliation with an aesthetics of transparency, and their potential use as surveillance mechanisms within an accepting population.

Sensoree's NEUROTiQ

According to the design company Sensoree, NEUROTiQ is "a brain animating fashion. A knitted 3-D printed EEG brain sensor that maps thoughts and exhibits brain states with color" (Sensoree 2014). More specifically, it translates signals from an Emotiv EEG headset onto the 3-D-printed knitted surface.[3] The headdress is very reminiscent of the EEG cap, but instead of feeding data to an oscilloscope or a computer that provides readouts in numeric or wave formations, the cap itself lights up based on the EEG data received from the small sampling of electrodes in the underlying headset.[4] Designed by a team at Sensoree that includes Kristin Neidlinger, Grant Patterson, and Nathan Tucker, the headpiece is intended to externalize the brain state of the wearer for all to see and to colorize the results for easy access to the wearer's state of mind (figure 1.1). The color-coding for NEUROTiQ is reminiscent of the mood rings of the 1970s, but instead of temperature, the scale is based on the frequencies of electrical potentials registered by the EEG wearable. The five colors correspond to established ranges of electrical activity (termed gamma, beta, alpha, theta, and delta waves), however, the color coding and translation of color to brain state arguably belie certain cultural assumptions about how and when the brain is "working."

Sensoree is the brainchild of Kristin Neidlinger, an MFA who built her enterprise around the idea of bio-therapeutics: technologies that can provide feedback to their users (and others around them) to relieve the symptoms of SPDs that range from autism to attention deficit–hyperactivity disorder. Previous to NEUROTiQ, Neidlinger created the GER (galvanic extimacy responder) Mood Sweater, FUR (personal space sensor), and the Inflatable Corset, among others. Each wearable is intended to "bring people back into their bodies" (Sensoree 2015) by making them—and sometimes those around them—aware of their mood and their environment. The FUR wearable senses movement behind you and "rises up and to warn and protect personal space dimensions" (Sensoree 2016) while the Inflatable Corset will increase pressure on your torso if it senses increased stress/arousal levels, ostensibly reminding you to calm down. In an early interview, Neidlinger explains that her dream for the company is to create a biofeedback therapy center to which individuals can come, be monitored, and leave with a customized wearable. Or, as one interviewer so eloquently puts it: Sensoree will be "like a spa that produces clothes for you" ("Artfuture" 2010).

Figure 1.1. NEUROTiQ EEG headdress. Design by Kristin Neidlinger ©SENSOREE; photo by Elena Kulikova. Used with permission.

Commentators believe that Neidlinger's wearables are unique in their packaging. In a review from 3DPrint.com, Eddie Krassenstein (2014) begins by setting a futuristic scene: "Imagine a time in the future when communication is not done through verbal speech, perhaps not even with hand motions or facial expressions. Imagine a time when we can speak to one another, no matter what our native languages are, via signals that our brains send to each other." His commentary indicates a desire for technologies to intervene in human communication—much as they always have—with the goal of cutting out problematic translations or stumbling blocks (language[s] or verbal communication). But Kassinger's commentary also draws our attention to design and the explicit claim that Neidlinger's NEUROTiQ is novel in the world of fashion. Neidlinger's fabricating technique is innovative, shedding the typical bulk of 3-D printing in favor of a knitted fabric embedded with light-responsive globules.

However, in its assumptions about human physiology, Neidlinger's NEUROTiQ follows nearly a century of making the body's processes visible and translatable. To this end, Sensoree's webpage includes quotes from various thinkers; two of them refer directly to the logics of psychophysiological communication:

> The body reveals what words cannot.—Martha Graham

> All ideas, all mental images, all emotions reveal themselves physically.
> —Augusto Boal

(SENSOREE 2015)

Graham, a dancer, and Boal, a chemical engineer turned actor, would be interested in the ways that bodies reveal the unseen and the unspoken. Taken in the context of Sensoree, and NEUROTiQ, their words evoke longer and more complex genealogies of psychophysiology: the idea that we can look to and through the body for physiological signs of psychological struggles—including arousal (often coded as stress or excitation). EEG wearables tend to embrace the machine as interlocutor for the self. Users are often portrayed as naïve about their own states of arousal, which are revealed to them, and their audiences, via an externalized display that can take several forms. In short, users are trained away from introspection and into technological mediation, a process that has been in the works at a professional level for nearly a century. In *Feeling Mediated* (2014), Brenton Malin argues that many psychologists stopped trusting introspection just as they began to develop machines that could measure physiological data that could then be translated into information about one's state of mind or emotional state. Machines became the mediators—a paradigm shift that happened in numerous fields, including medicine—at about the same time that we began to question things like memory, self-reporting, and subjectivity. Rather than encouraging individuals to look inward, or even to self-report, the mechanical report comes to dominate (Dror 1999, 2001).[5]

The trend away from introspection and toward mechanical reporting is both a product and producer of the "burdens of intimacy" (Malin 2014, 183), a burden that continues to inform psychophysiology—not to mention the EEG wearables market. On the one hand, early psychophysiological experimenters needed a way to distance themselves from their own subjectivity and the emotions of their subjects—to create an ostensibly more objective space within the laboratory. One means of creating this distance was by foisting the burdens of emotional absorption and translation onto the mechanical equipment in the laboratory. On the other hand, and as a result of this laboratory preparation, psychophysiological instruments have aided in the production of the burden of intimacy—particularly in the current, wearables market—by encouraging the externalization of feelings pertaining to attention, focus, or fatigue. Wearers are given a kind of amnesty for their internal states so long as they participate in reporting out.

In repackaging EEG as a fashionable headdress with a bio-therapeutic purpose, Neidlinger highlights, magnifies, and superimposes the data output from a typical EEG cap directly onto the user's head. In so doing, she reduces the space between data gathering and data translation, ostensibly collapsing the internal and the external, creating an object that exemplifies her conceptualization of extimacy and (perhaps inadvertently) demonstrating Rachel Hall's aesthetics of transparency. However, and within the larger history of psychophysiology, the "forced correspondence" between internal states and external recordings remains an ambiguous and ambitious undertaking. So while Neidlinger's NEUROTiQ makes explicit the burdens of intimacy, particularly for individuals with sensory disorders, the headdress also draws attention to the potential incongruity of data and interpretation. In this respect, NEUROTiQ is both a literal translation of brain wave ideologies and the perfect execution of an aesthetics of transparency in which everything and nothing are, at once, revealed.

NeuroSky's Necomimi

One of the more flamboyant, accessible, and attainable EEG-based wearables is Necomimi, which can be purchased on Amazon.com for just under $50 (figure 1.2). Relying on biomimicry, this wearable is essentially a set of fabricated ears affixed to an EEG headset and worn on the top of the head.[6] As its base, Necomimi uses the same Emotiv headset as NEUROTiQ; however, the Necomimi headset uses a single channel recorder and includes two "ear" buds that allow users to customize the way that the headsets look by using interchangeable ear covers. Several companies sell ear covers, including, most notably, Emoki. Like NEUROTiQ, these wearables are intended to be—at least partially—seen (though many young women reviewing the product on YouTube suggest elaborate ways to hide the actual headset under wigs, in hairstyles, or through the use of hairpins). The variety of available ear extensions runs the gamut from devil's horns to cat ears to just about any kind of furry cosplay one could imagine.

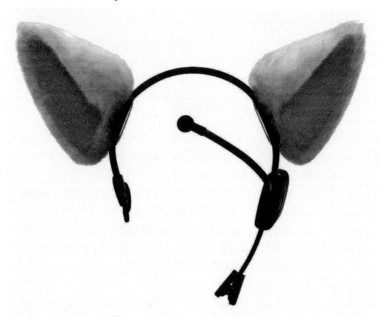

Figure 1.2. Necomimi Brainwave Cat Ears. NeuroSky, http://necomimi.com/Gallery .aspx. Used with permission.

Various websites that demonstrate Necomimi envision it largely as a party game or a novelty item. The company's website describes the product as follows: "'See What Makes Your Ears Wiggle': Show the world what's really on your mind and impress your friends with some of the most advanced brain-wave technology available! Necomimi's cat-like reactive movements show how interested or relaxed you are in real-time. It's a fun, quirky addition to parties, cosplay, bachelorette weekends and tailgating at your favorite sporting event. Anytime you want to entertain your friends and family, wear Neco-mimi!" (Necomimi 2015). Notably, the emphasis here is less about knowing the self and more about entertaining others. However, the copy goes on to explain the device's scientific basis and the "meditation" and "attention" algorithms on which its movements are based (figure 1.3).

Necomimi relies on an Emotiv headset, algorithms from NeuroSky, and the creative inspiration of Neurowear. Founded in 2011, Emotiv is a bioinformatics company that produces both hardware and software. NeuroSky has an established image as a fitness company for the body and mind; its EEG biosensor solutions have been adopted by several top-ranking universities. Neurowear is

How We Sense Your State of Mind

Step 1 Neurons firing in the brain give off electrical impulses, which are read by the forehead sensor.

Step 2 The Necomimi headset captures brainwave data, filters out electrical noise, and interprets your brainwaves with NeuroSky's Attention and Meditation algorithms.

Step 3 Your mental state is translated into ear movements and shared with those around you!

Figure 1.3. Brain wave sensing for the Necomimi. NeuroSky, http://necomimi .com/HowNecomimiWorks.aspx. Used with permission.

a "project team" based in Tokyo whose mission "is not the development of any specific technology. We are focused on creating what we call 'communication for the near future,' and we just leverage the power of technology to achieve that vision. For example, the neural sensing part of the Necomimi prototype is based on the Think Gear chip provided by our partner Neurosky Inc." (Neuro-wear 2015). As with NEUROTiQ, Necomimi is backed by at least one company interested in novel means of brain communication.

Necomimi (literally translated as "cat ears") has repackaged EEG as an animal prosthesis—a sixth sense, if you will. The plastic nubs at the top of the Emotiv headset do not serve any function as far as EEG data collection is concerned; they are an aesthetic choice intended to serve as a data output device. But why ears? At least as able-bodied humans experience them, ears are for data input: vibration transmuted to sound. As prostheses, the plastic nubs on Necomimi modify not only our ears (changing their assumed function) but also our brain, ostensibly connecting the two and allowing for more direct communication; their positioning implies that we need more sense and communication organs than we currently have. Expressing a similar sentiment, Kristin Neidlinger includes a quote from Robert Hooke on the Sensoree webpage:

> The next care to be taken, in respect of the Senses, is a supplying of their infirmities with Instruments, and as it were, the adding of artificial Organs to the natural . . . and as Glasses have highly promoted our seeing, so 'tis not improbable, but that there may be found many mechanical inventions to improve our other senses of hearing, smelling, tasting, and touching.—Robert Hooke
>
> (SENSOREE 2015)

Although the Hooke quote is from 1665, it appears repeatedly in much of the wearables literature and press, indicating that one goal of these bodily technologies is to recognize the senses we lack and enhance the senses we do possess. As prostheses, Necomimi's "ears" reveal certain assumptions about the human brain as an output device—one that emits signals that we cannot normally see or "read." "Prostheses imply certain enhanced knowledge of the bodily organ they modify" (Fortunati, Katz, and Riccini 2003, 16). Necomimi ostensibly makes it possible to sense and see aspects of our mental life that were invisible to the naked eye.

Simultaneously, and under the auspices of various animal ear covers, the ear buds evoke images of biomimicry. With names such as "Snow-Leopard" and "Jungle Leopard," one cannot help but think of Necomimi as tapping into our "animal instincts" and making (synthetic) "fur" fashionable once again. Cultural constructions of animals often figure them as instinctual and tied to their biology. Here, the animal prostheses allow users to explore a kind of sixth sense that appears to be inaccessible otherwise but that clues us into our animal instincts. The ear movements one sees are, or should be, familiar given that they resemble movements many of us have seen in animals, particularly pets—to which we often ascribe human emotions and states of mind. In the abstract, these ear movements may appear to make little sense, but in the context of arousal, they also appear to be more immediate and readable without the help of an expert or even basic instructions. If NEUROTiQ renders visible the burden of intimacy, Necomimi ascribes arousal to an animalistic unconscious that is, ironically, only accessible through technological intervention.

CRCMining's SmartCap

Beyond a prototype or a party favor, the SmartCap makes use of EEG for the purposes of occupational surveillance and safety. In 2008, Australian company CRCMining developed a hat prototype with an integrated EEG monitoring system intended to measure fatigue (figure 1.4). The cap has sensors embedded in the lining (yet remains washable!), and users simply put a card, termed a "SmartCap Fatigue Processor," into the bill of the hat. The sensors within the cap collect and measure dry EEG each second; the card translates the data and wirelessly sends them on to a central system. Within the user's vehicle the data also appear on a real-time display. The SmartCap was developed by Dr. Daniel Bongers as a way to minimize the number of fatigue-related injuries and deaths among miners who drive long hours, over the same routes, often in the dark of night. The cap was site tested in 2010, and by 2011 it was in use at various mines around the globe. The company now offers several styles, including a base-ball cap, a woolen hat, and a headband.

While the technology has prevented hundreds of fatalities, the basic idea behind it (measuring fatigue) is not new, nor is the use of EEG sensors.[7] Indeed, scientific studies using EEG to measure driver fatigue were reported in the *Chicago Daily Tribune* in the mid-1950s ("Measure Driver's Brain Waves" 1957). This time, however, the SmartCap overcame a bigger barrier than extra-lab mobility; it works because it is accepted—it fits within its industry's aesthetic. "With the look and feel of a typical baseball cap, the SmartCap has overcome

Figure 1.4. SmartCap baseball hat. SmartCap Technologies, www.smartcaptech .com/our-product/. Used with permission.

operator acceptance problems experienced at mining sites where camera or response based technologies have been implemented in the past" (Forster 2011). Or, as employees in advertising videos put it, "Since a lot of operators are already wearing baseball caps that lets us introduce this with minimal disruptions" (CRCMining 2015a). "In terms of fit, no issues; it just feels like I'm wearing a helmet or a normal cap" ("Putting a Cap on Fatigue" 2013). Both comfort and compliance are benefits of the SmartCap system. The latter is particularly important given that "the mining industry predominantly is male dominated; it's a very macho industry. The sign of admission to something is looked upon as a weakness. You don't want to tell someone you're tired because you're stronger, you're faster, you can work longer; but that sort of thing can lead to errors, it can lead to incidents, it can lead to injuries. It affects families; it affects more than just the people at the mine site" ("Putting a Cap on Fatigue" 2013). In this explanation, the mine site extends outward in a Venn diagram of ripples to larger units, such as the family and the greater mining community. The solution appears to be as simple as a baseball cap.[8]

CRCMining, the research conglomerate that created the SmartCap, is invested in bringing together university researchers, business experts, and mining industry leaders in order to create new solutions to problems in mining and other heavy industry. The collaboration of various industries allows for the pooling of research dollars and the sharing of other resources, both of which lead to the production of additional innovations more quickly: "The technology behind the SmartCap was developed within CRCMining, supported by four universities and 13 industry partners including equipment manufacturers and mining companies. Anglo American Metallurgical Coal (previously Anglo Coal Australia) and the Australian Coal Association Research Program (ACARP) supported this work for a number of years, which led to two successful field trials at central Queensland surface mining operations, where it was used by operators in haul trucks, excavators, dozers, graders and water trucks" (CRCMining 2015b). Primary credit for the SmartCap's development goes to Dr. Daniel Bongers, a native of Brisbane with a PhD in mechanical and space engineering and an artificial intelligence specialist. Dr. Bongers was even featured in an ad campaign for AngloAmerican, one of the mining companies that implemented the SmartCap. Here, the playful copy reads: "Dr. Bongers' Brainwave, a Cap That Measures Brainwaves." The smaller print reads: "At AngloAmerican safety comes first, second and third. So even if an idea at first seems a bit unusual, we take it seriously if we think it could help." The copy eloquently illustrates not only the company's desire to embrace the final, successful, product but also its attempts to valorize the—sometimes "unusual"—ideas that emerge in the process of design.

The SmartCap plugs drivers into a larger system of surveillance and control via its "Fatigue Manager supervisor interface," which "delivers audible and visual fatigue alarms based on site-specific criteria. It also allows review of single and multiple shift fatigue histories of individual users and also the remote customization of various fatigue alert criterion. This system works with the in-cab equipment to provide a comprehensive fatigue monitoring solution, which complements existing fatigue management strategies" (SmartCap 2015b). And—as we might suspect—the company has developed an entire discourse related to SmartCap use: from the "SmartCap Fatigue Manager supervisor interface" to the "SmartCap Fatigue Manager Server" to the "SmartCap Fatigue Processor," the circuit is a closed loop of branded products. And, like good old St. Nick, the SmartCap essentially knows when you are sleeping, it knows when you're awake, and it knows if you are trying to outsmart the system by not wearing the cap—yes, there is even a "cap off" alarm that will sound within fifteen seconds of a driver's noncompliance.

In this sense, as in Rachel Hall's original conception, wearing one's Smart-Cap means enacting "transparency chic"—the fashion of rendering oneself open to examination, a process that can help one accrue particular social or occupational status. The ideals of transparency extend even as far as SmartCap's client list. When asked in its FAQ for a list of corporate users, SmartCap's reply is loaded with expectations of translucence: "We respect the privacy of our clients, and will only share this sort of information if we have the express permission of a particular site. That said, we expect that a number of our current clients will soon make their experiences and future plans public" (SmartCap 2015a). The aesthetics of transparency that inform each of these EEG wearables are always already mediated in complex and simple ways, through the companies that employ the technologies, the designers of the black-boxed algorithms, and the wearers themselves.

Transparency Mediated: Translating Data; Designating Norms

Featured prominently on the NeuroSky home page is the slogan "Brainwaves. Not thoughts" (NeuroSky 2015a). For NeuroSky, the implication is that EEG wearables measure not the ephemeral, or even the psychological, but the material and the physiological. Its EEG wearables measure the concrete: the brain's electrical activity, not the mind's psychical machinations. The company's claim depends not only on the brain wave ideologies I outlined in the introduction but also on the assertion that the technologies necessary for instrumental intimacy are themselves transparent. However, if these wearables are measuring the brain's electrical activity, how are those brain waves related to the electrical potentials being collected and, in turn, being represented? How is data input translated into meaningful output? And what have these outputs been crafted to mean for different audiences? If transparency is always already mediated, then what about aesthetics of transparency?

How EEG wearables translate and display data has implications. Even beyond the often proprietary (and therefore largely unavailable) algorithms through which EEG data are parsed, the output made available to users shifts conceptions and expectations of brain states. Science and technology studies (STS) theorist Paula Gardner and biologist-turned-artist Britt Wray argue that "consumer-grade EEG *obscures* the work of capture, and processing, and prohibits synthesis beyond the narrow brain wave theory of cognition. In addition, packaging literature explaining the EEG devices unproblematically correlates brain waves to cognition, which is presented simply as the *effect* of the brain waves" (Gardner and Wray 2013).[9]

At stake are the (new) norms established by data visualizations that affect our understanding and definitions of arousal, often expressed as varying levels of attention, focus, and fatigue. In psychophysiology and neuroscience, definitions of "arousal" vary. Sometimes arousal is linked to wakefulness and consciousness (Laureys et al. 2009; Schiff and Plum 2000); sometimes it is used interchangeably with "activation" (VaezMousavi et al. 2007), which one often hears in conjunction with brain visualization technologies such as fMRI, PET, or MEG. For our purposes, Mohammed VaezMousavi et al.'s distinction between the two terms is useful: " 'arousal' refers to the individual's energetic state at any moment. . . . The task-related 'activation' is defined as the change in arousal from resting baseline to the task. During a task, the current (activated) level of arousal affects physiological response amplitudes, while the task-related activation affects behavior/performance on the task" (181).[10] In other words, arousal is a measure of energy levels at a specific moment, whereas activation is the change in energy levels (arousal) measured as one focuses on a specific task. This distinction is important because, as VaezMousavi et al. go on to argue, performance within a particular situation is not linked to arousal but to activation, or the change in arousal from one state to another. Algorithms that link the two depend on multiple factors, including the objectives of an experimental study, variations in stimuli, issues of statistical significance, and so on.

EEG wearables typically depend on both activation and arousal, but they do not always mark the difference between the two. Moreover, the fact that we are arguing about language, definitions, and meaning indicates that there are gaps between what we desire to represent and how that representation is enacted. Or, as VaezMousavi et al. maintain, even as they intend to stake their claim about the use of specific terms, "various terminologies that have been used to describe states of attentiveness in the CNS include arousal, alertness, vigilance, and attention. As most terms are used extensively with diverse associations, it seems that none are ideal to describe these cortical states" (2007, 180).

NEUROTiQ, Necomimi, and SmartCap collect, interpret, and translate particular ranges of EEG data into different quantifiers or outputs; in so doing, each of these wearables limits the range of electrical potentials that can be expressed. In a technology like SmartCap, electrodes are set up to measure "fatigue," a particular state that has been defined by several research groups (including A. Craig et al. 2012). Fatigue levels are displayed to drivers and their supervisors through a small dashboard monitor that displays a number between one and four. NEUROTiQ relies on a different mix of potentials read through different algorithms to represent five brain states, described as "Deep Sleep," "Meditative," "Relaxed," "Alert," and "Multi-Sensory"; each of these interpreted states is then translated once more into a color. Necomimi, which records only a single channel of EEG data, claims to measure three things: alertness, interest, and attention; these states of arousal are then translated into prosthetic ear movement.

The myriad possibilities for data visualization in EEG wearables should remind us not only that data representations are fungible but that translated representations fundamentally change data's power and meaning—particularly for lay audiences. We have seen evidence of this time and again from anatomical representations (Waldby 2000) to PET and fMRI scans (Beaulieu 2002; Dumit 2004) to astronomical images of nebulae (Greenberg 2004). In the case of PET brain imaging and its use in the courtroom, Dumit explores the power and limitations of the visual: "One cannot, for instance, actually see mental illness in the brain, one can only see the large variations in different brains and attempt to correlate certain kinds of brains with certain diagnoses of persons—as normal, schizophrenic, depressed, and so on. The desire of course is for the machine-imaged brain to replace the psychiatrically-diagnosed mind, the 'holy grail' of biological psychiatry" (Dumit 1999, 182). In EEG wearables, we find evidence of a similar desire: that the objective measurement of "Brainwaves. Not thoughts" can bolster the illusion of a direct connection to states of mind and arousal.

In EEG wearables, the use of color, animal ear movements, and simple four-point numeric scales ostensibly allows the data to be transparent, to "speak for themselves"—these visualizations make it seem as though the data are so simple that these translations capture what is collected without leaving a huge remainder unanalyzed or unaccounted for. Or, put another way, the visualizations of EEG wearables offer the illusion that what you see is what you get.[11] Representations of data imply that there are algorithms that can separate out particular electrical activity as dominant or pronounced, leaving other brain activity out of the picture, and allowing the data to translate directly into "active" or "in the zone" or "relaxed." Yet we know all too well that the expert's voice and selections are built into the wearables themselves, foregrounding certain data and backgrounding other data. For example, the various headsets employed as foundations for EEG wearables collect electrical potentials on one to fourteen channels that may or may not correspond to the states of arousal that are supposedly being displayed. As Mihajlović et al. assert concerning the headsets on which the wearables are based,

> due to the inherent difficulty in designing a complete EEG solution, most non-clinical EEG solutions are designed for general purpose EEG applications with a lack of support for sophisticated signal processing and effective feedback generation. They suffer from a number of forced tradeoffs that severely limit the usefulness of such systems. For example, Emotiv Epoch has a suboptimal design for a number of applications, given the fixed electrode position, which is not according to the 10–20 system and has suboptimal headset design. . . . This is partially compensated by offering software development kits for implementing signal processing algorithms. The solution offered by NeuroSky is oversimplified in the design such that it can only be used for toy applications in a niche market (i.e., a small set of serious games). To avoid the signal quality evaluation issues, it can only be used in specific set[s] of games for which signal processing and interface design is provided. (Mihajlović et al. 2015, 17)

Pulling back from the specific limitations of Emotiv's hardware or NeuroSky's software, it is important to recognize that "currently, no EEG solutions available on the market are designed with a specific EEG application in mind" (17). The interpretive work, then, must come in the translations and visualizations of various data for a particular user.

NEUROTiQ and the SmartCap sit at very different ends of the wearables spectrum. NEUROTiQ is a high fashion one-off that is unavailable on the consumer market, but it emerges from a longer line of products dedicated to sensory processing disorders. The SmartCap, by contrast, is a widely available and implemented wearable that has real-world consequences including job-related performance evaluation. Both can help us explore issues of translation.

First, how we talk about brain states, and which brain states we *want* to experience, affects the power and type of translation employed. Imagine the party envisioned by that 1970s Australian advertisement in which the wearer looks to their mood ring for information about herself. One possible response to the question of "How are you?" is "I'm marvelously blue, thank you'" ("Changes Colors" 1970). In that case, the translation of temperature to color to mood matters; it tells us a lot about the ways that emotions are normed and assigned particular alter-egos in colors or numbers. For this user, being "blue" does not imply depression; instead, "blue" and being blue represents something wonderful: she is "marvelously blue" because that is what the device has authorized as the correct translation of data to interpretation via the medium of color.

Likewise, NEUROTiQ has five possible color states linked to the five identified types of brain wave activity. Each kind of wave has been assigned a title and a color. Like the mood rings of the 1970s, the "Emotive Display" lights up differently depending on the types of electrical activity being emitted by the user (figure 1.5). These might qualify as "honest signals" ("Using Wearable Technology" 2013) because they are outside of the range of conscious control; however, these particular states are somewhat easier to control than one's temperature by, for example, simply closing one's eyes to move from beta to alpha waves. Given the way that the headdress is designed, it is difficult to imagine that one would see red ("deep sleep") very often or that one would not interpret most of the activity as gamma, represented as a mash-up of other waves and sensory stimuli. The most likely colors are blues and yellows. And these color choices matter. The designer of NEUROTiQ mentions in one interview that when she tested another similarly color-coded device, the Mood Sweater, in different countries, the users tried to keep their activity within culturally desirable spectrums—and sometimes those spectrums differed from

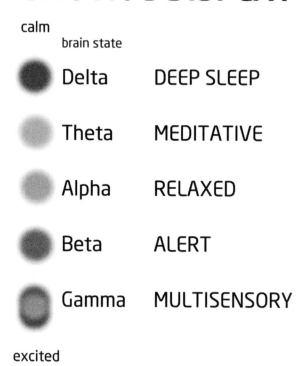

EMOTIVE DISPLAY

calm
  brain state

Delta   DEEP SLEEP

Theta   MEDITATIVE

Alpha   RELAXED

Beta   ALERT

Gamma   MULTISENSORY

excited

Figure 1.5. NEUROTiQ emotive display. Design by Kristin Neidlinger ©SENSOREE; photo by Elena Kulikova. Used with permission.

the designer's goals. In Brazil, for example, users "wanted to be pink and red all the time; they wanted to be excited, in love" ("Wearables+Tech" 2014), not in the relaxed "blue" zone that the designer had assumed to be a desirable, normal state.[12]

If NEUROTiQ displays what is "right" with you while also helping to situate you within a matrix of normal or expected responses, the SmartCap focuses on what is "wrong" with you: the moments when your body—and therefore your state of fatigue or arousal—fails to comply with an accepted standard. The SmartCap provides a readout in numeric form, a scale from one to four, which besides being displayed on a screen in the cab, is available for supervisors to observe (figure 1.6).[13] A simple number indicates how fatigued an operator

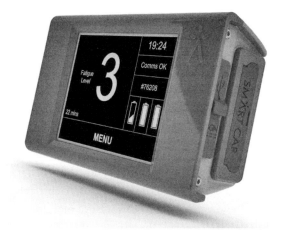

Figure 1.6. SmartCap dashboard display. SmartCap Technologies, www.smartcaptech .com/our-product/. Used with permission.

is. All pretense of color and animal movements are absent; indeed, it does not *seem* like much of a translation at all—particularly because it is also only measuring one thing: physiological fatigue. But it is a translation.

First, this definition of fatigue does not take into account emotional fatigue or the fatigue one might feel from working a routine or mundane job day in and day out. These alternate definitions help to situate the specific kind of fatigue that is assumed by this wearable. Second, although it remains proprietary, the company claims that it uses a special algorithm to measure the fatigue of the driver. In this respect, the wearable and its visual display are both black-boxed technologies. All that users need to understand is that if the numeric readout creeps toward a four and remains there for any period of time, they can be pulled off the workforce and asked to rest in a quiet room for up to an hour before being allowed to go back to work. The driver, who may or may not have felt in touch with his or her fatigue level—"you think you've had a good sleep, you think you're relaxed" ("Putting a Cap on Fatigue" 2013)—has been overridden by the machine. The small range of results make the dangers of fatigue seem close and tangible—only a few shifting numbers away. Moreover, the fact that the number scale goes up—a four being the most fatigued, matters; it defines the problem of fatigue as an escalation, a progression that could be stopped. Fatigue has been redefined by this EEG wearable as a simple numeric score and as "a form of impairment" ("Coal & Allied Fighting Fatigue" 2014). As with the other wearables, drivers might be able to train themselves to register a more alert number (though this remains unclear given that we cannot access the algorithm).

It should come as no surprise that the translation of EEG data matters. Visualizations of data are varied and tell us much about the systems of *mental* dispositions that inform the design and capacities of EEG wearables. The claim may be that we are seeing "Brainwaves. Not thoughts." Yet we are actually seeing neither; instead, we are seeing a designer's representation of activity and arousal. The instrumental outputs are too muddied for designers, users, or audiences to parse whether that arousal comes from electrical potentials in the brain, overhanging power lines, heartbeats, or eye movements. But what we *see* matters, and EEG wearables certainly give us public displays of arousal.

EEG Wearables' Speculative Futures

Public emotional displays were once a common feature of the American social landscape: from love letters addressed to family members of a potential spouse to particular displays of affection, the emotional landscape was rich and varied and externalized (Malin 2014). With the advent of laboratory research into emotion and the rise of mass media technologies that ostensibly stimulated emotions (sometimes in dangerous ways), emotional displays began to retreat, and machines began to play a role as intermediaries between people, their emotions, and those who might be affected by such displays (Dror 1999, 2001; Malin 2014; Stearns 1994). This rich and varied history of externalized emotion continues to take various turns in and out of fashion. In the 1970s, mood rings represented the latest in fashionable feelings. In the early twenty-first century, wearable EEG technologies have made public displays fashionable once again—but this time, the display has been redefined not as amorous emotion, or dangerous subjectivity, but as mental arousal.

Looking to technology for a sense of arousal or a state of mind has arguably rendered us more alien from ourselves; this way of looking also reveals our *selves* as permeable, fractured, and ever open to being reconfigured by interactions with technologies. Once we defer to machines and begin to think of transparency as a desirable aesthetic, it becomes easier and more likely to develop instruments—and wearables—to satisfy a transparent chic. Turning to machines means that even in cases where we are sharing our states of mind socially, we are not responsible for that sharing—it is out of our control—and those who are listening to or receiving our signals need not care about the underlying causes. States of mind, sensations, and fatigue all become data points rather than actual ontologies to be inhabited, discussed, understood, or analyzed; they are read-outs on a machine interface to be dealt with according to a prescribed plan. For example, if your fatigue levels are high there is no need to discuss how you feel; instead, just follow protocol and sit in a quiet room. Moreover, the promised technological revelations are potentially hyperbolic. The ears of the Necomimi do not divulge how one is feeling; the ears offer up only those states of mind valued by a particular designer for and within a particular culture at a given point in time. In the case of these EEG wearables, we are most interested in alertness, relaxation, and relative levels of stress and fatigue. These wearables do not register sadness or discomfort or anger—or a myriad of emotions, feelings, or states of mind deemed uninteresting or unmeasurable.

While we do not yet live in a society in which everyone is wearing animal ears or 3-D-printed light-emitting globules on their heads, we do live in a world of SmartCaps and Fitbits and Apple Watches that promise to monitor, record, and provide feedback about many aspects of our physiology—things that we have been taught to control and things that we never imagined having control over. In each case, popular culture and technology have merged—once again—to make fashionable certain kinds of revelations about the self. Largely, these have to do with health and fitness, but as we gain access to monitoring systems based in EEG, we may problematically assume that we are moving closer to the center—to the organ actually responsible for the regulation of our bodies and, by extension, ourselves. Who knows what kinds of bodily processes will become fashionable to display, to regulate, to attempt to hide. Wearing our brain waves on our sleeves—or heads—may sound like futuristic thinking right now, but the consequences could be unnerving (no pun intended). These are some of the questions that animate subsequent chapters; for now, it is enough to imagine something like a SmartCap in everyday use: in schools, at work, at home, at the doctor's office, at the DMV. Only it's not *just* measuring fatigue anymore. You may or may not have access to the data; you may be judged on your brain's ability to interface with or control the technology; you may accept this as a perfectly fashionable thing to do.

2

In the Zone

Better Training through Alpha and Beta Waves

> Change your brain waves; change your life.
> —BIOCYBERNAUT INSTITUTE

In July 2006, with the World Cup finale on the horizon, a Canadian newspaper ran the headline "Italy's Weapon Is All in Their Heads" (2006). Four of AC Milan's players had been training, not only on the pitch but also in what has been dubbed the "Mindroom"—a bio- and neurofeedback laboratory in Italy. "In the Mind Room, the athletes lie on reclining chairs, their bodies strapped to the ProComp device that measures seven physiological signs—from their brain waves and muscle tension to their breathing and heart rate." While it may seem counterintuitive that athletic training could depend on relative inactivity, this example illustrates that physiological signals from muscle tension to electrical brain activity can be subjected to a specific kind of exercise: bio- and neurofeedback. Neurofeedback is a form of biofeedback; in its most basic form, neurofeedback converts information about the electrical activity of the brain into sensory information (or feedback). Subjects are expected to use this information to train their brains into particular patterns of activation.[1]

It is important to note a distinction between *neurofeedback* and the larger, umbrella term *brain training*. The latter is often attached to work or games that encourage mental cogitation in the form of word and number puzzles and memory tasks. Brain training does not always involve EEG hardware, yet the discourses surrounding it are often similar to those that inform neurofeedback—at least in its most recent formulation.

Contemporary neurofeedback is both product and producer of shifting definitions of conscious control. Through the quantification of electrical activation in the brain—and new interpretations, expectations, and applications for these data—we are offered literal and discursive control of our most central (and once presumed to be *inaccessible*) organ, our brain. More specifically, through neurofeedback, researchers have begun to quantify states that were once intangible, for example, being "in the zone" or achieving mastery. Whether the quantification process or the results are accurate, the assumption that we can voluntarily enter and exit these desired states relies on certain developments, including theories of plasticity and conscious control. In this chapter, I focus on two particular neurofeedback approaches: exercise, represented by mental gymnasiums such as the Mindroom, and accelerated learning, represented by Chris Berka's company, Advanced Brain Monitoring. Both rely on laboratory-grade EEG wearables that allow for participant mobility.

Elite sports is not the only market for neurofeedback, nor is sporting the only place in which these discourses thrive. Indeed, athletes are only one of a dozen target markets. Take, for example, the Biocybernaut Institute, a set of training centers for business executives and other professionals who want to improve their workplace effectiveness.[2] Its slogan, "Change your brain waves; change your life," offers a window into the company's—and our larger culture's—discursive claims concerning brain wave ideologies and neuroscientfic control. Indeed, the institute promises that its customers will gain control of their bodily systems: "The term Biocybernaut derives from its three linguistic components. 'Naut' is an Ancient greek word meaning voyager. Cyber derives from cybernetics; which is the study of control systems. Bio is meant to refer to an individual's biology and physiology. Hence, a Biocybernaut is one who journeys through his own biological control systems. At Biocybernaut, trainees learn to control their own brain waves" (Biocybernaut Institute 2015). Here, the slogan "Change your brain waves, change your life" becomes the responsibility of each cybernaut, who, like the *Fantastic Voyage* scientists, must enter and move through their own "biological control systems" in order to fix brain systems that are out of control. While economically inaccessible to the average person, the Biocybernaut Institute capitalizes on the ubiquitous brain wave ideologies I identified in the introduction: specifically, that our brains are electrically accessible and flexible.[3] As is evident from places like the Biocybernaut Institute, the metaphors and discourses informing neurofeedback training for elite athletes are part and parcel of our larger cultural milieu, one that includes consumer-based EEG wearables such as Muse, Mindo, and Lotus (aka Puzzle-box Bloom).

Several scholarly conversations are relevant to these inquiries: Jonna Brenninkmeijer's work on neurofeedback and conceptualizations of the self (2010, 2013a, 2013b), David Linden's work on brain control (2014), Brad Millington's arguments concerning brain training (2011, 2014), and Nikolas Rose and Joelle

Abi-Rached's work on the sciences of brain management (2013). These scholars all have contributed to a larger dialogue about neuroscience's effects on conceptualizations of selfhood and responsibility. We can build on their foundations to trace a broader shift in assumptions about conscious control and persistent metaphors of muscularity in and of the brain. By examining EEG specifically in the context of neurofeedback for elite athletes, I extend Brenninkmeijer's and Linden's work into the realm of athletics, augment some of Rose and Abi-Rached's arguments concerning the neuro, and recenter some of Millington's concerns about brain training around the specifics of athletic neurofeedback.

I begin this chapter by explicating a shift in our characterization of brain activation—from intangible, unconscious processes to a quantifiable signal deemed to be under our conscious control—that has influenced the current rise of neurofeedback in sports and beyond. Next, I examine two discourses about training involved in neurofeedback: exercise and accelerated learning. Each illuminates the literal and discursive representation of neurofeedback as an important tactic used to master the electrical activity in the brain, giving us a new sense of control and responsibility over what was once considered to be an inaccessible, unconscious set of physiological processes. Finally, I discuss discourses of mastery and their impact on neurofeedback regimes.

Creating Conscious Control: Neurofeedback, Plasticity, and Risk

Depending on the historical and cultural moment, scientists and laypeople alike have crafted particular assumptions about what is and what is not under our conscious control. These assumptions change over time as technologies render various aspects of our physiology more visible. Take, for example, the constructed division between our autonomic and central nervous systems. In neuroscience and physiology, scientists formulated a division between the two that was partially based on access and conscious control. Autonomic functions typically include heart rate, respiration, blood pressure, and skin conductance (sweat levels); aspects of the central nervous system are also assumed to be under less-conscious control, such as blood flow to areas of the brain and electrical activity in the brain. The division between the two systems allowed for the birth of technologies such as the polygraph, which relies on assumptions about the autonomic nervous system giving away information that is out of the subject's control (Littlefield 2009, 2011), but it also set the stage for potential interventions. In their discussions of the autonomic nervous system, Nikolas

Rose and Joelle Abi-Rached argue that "it is not just that these and many other processes essential to survival happen outside of our consciousness. It is that they are *in principle* unavailable to consciousness in normal circumstances, although some can become amenable to conscious control under particular conditions or with appropriate training" (Rose and Abi-Rached 2013, 219). This latter shift is key to the emergence of neurofeedback in the 1950s and the popularity of brain training and neurofeedback in recent years.

The brain's electrical activity has historically fallen into the categories of involuntary and unconscious. As David Linden characterizes the era of biofeedback's emergence, "The pioneers of operant conditioning of physiological responses in the 1950s and 1960s faced an uphill struggle because the standard view was that only voluntary motor acts could be brought under operant control, whereas autonomic responses—as implied by the term—could only be modified by classical conditioning" (2014, 17). When EEG neurofeedback studies began to appear in the popular press in the 1960s and 1970s, public perceptions of brain activation began to shift. A *New York Times* article from 1971, titled "Mind over Body, Mind over Mind," notes that

> until very recently, we assumed we were at the mercy of our involuntary nervous system. This notion helps to shape the self-concept most of us have, and it has been encouraged by medicine and psychology. Physiologists a decade ago believed that we were regulated by two distinct nervous systems: one was the great outer brain that gave us voluntary control of muscles through tentacles of nerves that travel down the center of the spinal column. The other, lower brain, is called "autonomic" or "vegetative." Its nerve fibers travel down the sides of the spinal cord and regulate survival functions, in the viscera, the heart, through emotional responses. These functions were held to be involuntary throughout the West—until the advent of biofeedback. (Luce and Pepper 1971, 35)

In this formulation, the distinction between the central and autonomic nervous systems is retained, but the accessibility of the latter has shifted.

Joe Kamiya's experiments, which focused on subjects' controlled reproduction and suppression of alpha waves, were some of the first to gain attention in the popular press. The publication of his essay "Conscious Control of Brain Waves" (1968) in *Psychology Today* drew national media attention. In an article for the *Los Angeles Times*, Kamiya is called a "pioneer among some 300 scientists who are conducting experiments with delicate electronic machinery and finding that we can develop much more conscious control over our mind and body states than most of us would have believed possible" (Gustaitis 1971, 6). In his popular retrospective, Jim Robbins likewise lauds Kamiya's work as "the first controlled experiment to demonstrate that brain waves, normally thought of as involuntary, were subject to voluntary control" (2008, 55). The popular press defined control somewhat vaguely: "After several sessions with the EEG equipment and the tone signal [haptic feedback], many people can continue to turn alpha up or down on their own, Kamiya has found. The extent of their control varies widely, as does the amount of alpha in peoples' EEGs" (Gustaitis 1971, 6). Nonetheless, revisions to conscious control were becoming legible to lay audiences.

Although Kamiya's research fell out of public view, his work brought advances in neuroscience and neurofeedback to the fore, and began the process of changing our minds about brain control. Take, for example, an article from the *New York Times*, "Can Man Control His Mind?" (Smith 1972). The article not only catalogues the biofeedback movement but also focuses on EEG as new form of biofeedback that is gaining prominence. We now have the Biocybernaut Institute, but in the 1970s, the Mind Control Institute was serving businessmen who wanted to gain the competitive edge EEG was offering: "The key to biofeedback training is teaching a person to recognize and then control his internal body rhythms, particularly brain waves" (F3). The article notes a certain distaste for the language of "mind control" among the biofeedback community. As a counter to these and other *Manchurian Candidate*–type fears, the article cites a Dr. Green who notes that the techniques are not hypnotic but "hypaotic" and acknowledges the importance of mechanistic feedback; "he also contends that the course should not claim to train alpha state unless it has a machine that can identify the state for students" (F4). In this story, echoes of instrumental intimacy aid in the legitimation of the 1970s' newest biofeedback techniques, but also notable are the shifting discourses of brain training and (mind) control.

Ideas and ideals about control continue to inform both popular and scientific neurofeedback literature, making neurofeedback a technique of control, both literally and discursively. In his popular monograph, *A Symphony in the Brain*, Jim Robbins notes, "We 'own' our central nervous system to a far greater degree than we imagine. . . . Neurofeedback shows us how powerful we are . . . [A]s biofeedback evolves, the 'aha' moment will come when we as a culture realize we have a great deal of control over our nervous system and accept that responsibility" (2008, xv). Jonna Brenninkmeijer's ethnographic research revealed this gem from "one of the neurofeedback course participants, a neuropsychologist," who "explains that he is interested in neurofeedback because it provides a solution, instead of only an identification of the problem. He describes colleagues who simply state that 'broken brains cannot be fixed,' while the great benefit of neurofeedback is that it 'gives control back to the client'" (2016, 125). For Nikolas Rose and Joelle Abi-Rached, "It is not that you have *become* your brain, or that you are identical with your brain, but you can act on your brain, even if that brain is not directly available to consciousness, and in so acting, you can improve yourself—not as a brain, but as a person" (2013, 223). The electrical activity that was once outside of our conscious purview has become something that we can—and should—act on.

Because we have increasingly sophisticated ways to measure electrical brain activity, and because we have characterized particular types of electrical activity as desirable, we now have new aspects of ourselves to monitor, worry over, master, and—perhaps, enhance. Moreover, within a neoliberal model, we are responsible for fixing aspects of ourselves that experts deem to be in deficit.[4] Ironically, this trend tends to manifest in discourses that advocate taking back control of our own brains (out of the hands of the experts). As one of Brenninkmeijer's interviewees notes, "We got in some kind of mind set in which we handed over our responsibility to the experts. We don't take responsibility ourselves; we go to a doctor and take medication rather than change ourselves. And I have a feeling that this might be going to change. I think there is something in the air, there is a shift going on. That people want to take responsibility for themselves" (Brenninkmeijer 2013b, 89). The paradox is that individuals are expected to take back a control that they may never have lost: in a historical sense, the brain's electrical activity was simply made visible for the first time; in a contemporary sense, the brain has now become available for neoliberal incursions.

One of the paradoxes is that shifting definitions of conscious control may or may not indicate real—or ultimately desirable—possibilities for training and self-regulation. Just because particular electrical activity in the brain becomes visible, chartable, and (potentially) changeable, how do we know we are looking in the right place? How do we know that what neurofeedback asks of us is actually what we want to achieve or accomplish or change? Neurofeedback may have grown in prestige since its early days, and is now served by several professional organizations, yet scientists still debate the electrical function of the brain.[5] "Despite all of the research that has since been done on the brain, no one is sure exactly what functions the electrical activity represents" (Robbins 2008, 19).

Some of the uncertainty surrounding neurofeedback derives from the brain's curious and liminal position: first, it is an organ, not a muscle, and as such occupies a paradoxical position in our discourses about the body. For many of our organs, it is difficult to imagine any kind of exercise beyond that old adage about a weekend bender giving one's liver a workout. However, for some organs, such as the lungs and heart, we have been prescribed a barrage of exercises that could strengthen their capacity and output. Second, like most organs, the brain seems both out of and under our control. It is the center of the central nervous system, yet many of its functions are controlled by the autonomic nervous system. Through meditation and other practices, people have claimed to more fully control organ function, but few of us have the experience to, say, consciously release bile from our gall bladder. Because it concerns an organ in this liminal position, neurofeedback must straddle the line between conscious and unconscious, physical and mental.

The very descriptions and practices of neurofeedback best represent this paradoxical dilemma, as they note both the conscious and necessarily unconscious aspects of the process. Neurofeedback typically works by converting the brain's electrical activity into sensory data that can be shared with the user via haptic feedback—a sound, a light, a computer game, or an image on a computer screen that illuminates or changes when the user has produced the desired electrical activation. The necessity of haptic sensors implies that we are not able to readily access or control our brain's electrical activity, and that learning to do so means relying on externalized conversions of internal states. As David Linden explains regarding the larger field of biofeedback, "Changes in heart rate, blood pressure or skin conductance, when they are subtle, are not obvious to our senses. For feedback training the experimenter thus needs to transform them into a sensory signal, for example, playing a tone for every heartbeat" (2014, 15). And when experts come between trainee and machine, an even more acute sense of procedural alienation emerges. Brenninkmeijer observed through first-hand experience that "during the trainings, I mostly felt uncomfortable because the question of what the practitioner and his computer were actually doing with my brain—and which effects this could have—kept on haunting me" (2013b, 46). Her reservations may be part and parcel of the contradiction at the heart of neurofeedback training: that no matter how we quantify the intangible, at the end of the day, even EEG training is often represented as out of our control. "Brain wave training is considered much more automatic," notes Jim Robbins. "A person does a forty-five minute session. After they leave, it's over. Nothing to be mindful of, no blue dots. Even though the client is changing his or her own brain, the training is so powerful and simple that it is very different from other kinds of biofeedback" (2008, 63). Thanks to the paradoxical nature of neurofeedback training and shifting discourses of conscious control, the brain is in a perfectly liminal position: it remains out of our control, at least until we *learn* to master it.[6]

As I discussed in the introduction, ideas of brain control depend, at least in part, on reconsiderations of the brain's infrastructure and physiology. We have moved from believing that the brain is relatively fixed—particularly after a certain age—to believing that the brain is actually less localized and more networked; for example, if one area of the brain is damaged, other areas might take over the tasks associated with that part of the brain. This finding (from lesion studies) has led to wider ideas about changeability in and of the brain. If the damaged brain can be modified or modify itself, then perhaps all brains can undergo more purposeful or directed modification. "A few decades ago, the idea emerged that the brain was plastic and malleable, instead of stable and immutable as most neurologists thought before. . . . This change in perspective brought the promise of brain intervention as therapeutic intervention rapidly into action, since the brain was no longer seen as an organ that determined behavior, but also as an organ that could be trained or enhanced to change this behavior" (Brenninkmeijer 2013b, 8–9). Thus, the idea of training the brain was born. "The arrival of brain training as a health pursuit is based mainly on recent neuroscientific findings that the brain is less like a blank slate or a computer-processing center (as metaphors of old would have it) and more like a muscle that can undergo atrophy or hypertrophy depending on its stimulation" (Millington 2014, 495). In EEG neurofeedback training, the basic idea is to change the electrical activation of the brain in response to particular stimuli. For example, athletes using a facility such as the Mindroom, might watch tape over and over again while receiving neurofeedback, until their brains produce the desired alpha or beta waves.

Machine and computer metaphors still rule, but descriptions of controlling the brain through neurofeedback have become more organic, relying on metaphors of a wilderness in need of maintenance or a chaotic space in need of structure.[7] Researchers talk about this process as pruning irrelevant connections or needing to find a technology that "supports, strengthens, and streamlines [the desired] pathway" ("Discovery Science's" 2013). Other descriptions of neuro-feedback characterize the brain as an unstewarded wilderness. Jonna Bren-ninkmeijer's dissertation (2013b)—and resulting book project (2016)—illustrate how some practitioners speak about neurofeedback. In one example, Bren-ninkmeijer attends a course for neurofeedback practitioners and records the following statement from the course supervisor: "It is like a field of weeds. If you walk through it once, and a few weeks later you do it again, you would take a different route. But if you do it very shortly after the first walk, you would take the same way. And like this you can create a path" (2013b, 81). The brain is imagined as a field of weeds—uncultivated—and rarely traversed in a purpose-ful fashion. This kind of imagining reinforces the idea that we are out of con-trol and must bushwack into existence the paths that will allow us to (re)gain some kind of control over our neurophysiology. Similarly, Jim Robbins de-scribes neurofeedback as taming potential chaos: "The neurofeedback model holds that the brain wave training increases the stability of that area of the brain a well as its flexibility, or its ability to move between mental states (from sleep to consciousness or arousal to relaxation, for example). It allows the players in the orchestra to play their parts better, to find the correct tempo, to come in on time, and to stop playing when they aren't needed" (2008, 45). In this passage, the gymnastic discourse of training, flexibility, and movement precedes the conductorless orchestra that needs to learn how to play in better harmony.

The conundrum with neurofeedback is not whether something measur-able is happening but whether that measureable change is having any specific effect. As Brenninkmeijer puts it, "What someone exactly has to do to change his or her brain waves, however, remains unclear" (2013b, 85). In line with Brenninkmeijer's questions, we might look to Pierre Beauchamp's description of neurofeedback. Beauchamp, the originator of the Mindroom, considers neuro-feedback to be part of a larger biofeedback program. In an introductory video for his program, Beauchamp assesses how and whether the technique works by arguing that

> biofeedback is taking the physiology and the EEG and putting it together and showing to the athletes what's happening in real time. And so the Thought Tech-nology equivalent that we're using, we take it a step at a time in our program. We start with muscle tension, respiration, heart rate, skin conductance and we put all of that together and we train each one of them and at the end the athlete under-stands which modality that they're using and can actually control that. The long-

term effect of that is increased confidence. The more the athletes relax the more likely peak performance is likely to occur. What we want is if we're looking at his threshold for performance and he's continually on the podium then we know that we're achieving our goal. (Mindroom 2015b)

Beauchamp does not name a specific threshold for determining whether neuro- (or bio-) feedback is working. The quantitative aspect of the method seems to falter under this kind of strain, yet he does maintain a metric that is measurable: whether the athlete is "continually on the podium." Whether this sustained success is attributable only to neuro- or biofeedback remains unclear, but for Beauchamp and other practitioners the results speak for themselves.

If aspects of brains are now deemed to be under our conscious control, and if theories of plasticity enable us to believe in the possibility of affecting change, then what remains is a motivating factor; this is where discourses of risk come into play. Even if we are unsure of the correspondence between neurofeed- back and brain or behavioral changes, we are often told that our brains, like our hearts, are at risk if we do not train them, keep them active, and use them to our advantage when solving performance or behavioral problems. Risk dis- course thrives, in part, on the quantification, measurement, and standardization of bodies. Nikolas Rose describes a "social history of numbers," noting among myriad examples that "expectations and beliefs are embodied in the framing of statistical inquiry, for example, in the form of explicit or implicit theories shaping what is counted and how it is to be counted" (1999, 204–5). Brad Mill- ington uses Rose's concept to illustrate the centrality of numbers "to our under- standing of 'healthy' and 'at risk' subjectivities" (Millington 2011, 439), as quantified norms also aid in the social construction of risk. Brenninkmeijer does a brilliant job of explaining the Foucauldian creation of brain clients, those who believe they are at risk and that the seat of their problems lies in their head: "Defining the problem must be followed by defining the problem as a brain problem. Trying to make people aware of their brains is becoming a key part of scientific and popular scientific discourse, and neurofeedback practitioners contribute where they can" (2013b, 88). Within this system, "people increasingly learn that their daily life problems are brain problems and from this point of view it is not surprising that some of them start to experi- ment with manipulating their brain" (50). If one does not work on one's brain, it has the potential to atrophy or deteriorate; only through training can one maintain some of the brain's robustness. This paradigm assumes that if someone is not taking advantage of available remedies for, now simplified and identified, brain problems, then one is at fault or is responsible for any negative results.

Alongside changing practices of neurofeedback and discourses of plasticity, risk provides the final linchpin in the triad of neuroscientific control. We have not only the ability but also the *responsibility* to intervene in our brain's electrical activity. What, then, are the available modes and methods for training the brain? How have muscular discourses affected the active training and mastery practices available through neurofeedback training centers and EEG wearables?

Exercising a New Muscle

Thanks to the ubiquity of EEG wearables, neurofeedback is undergoing a resurgence as both a product and producer of shifts in paradigms of conscious control and discourses of plasticity and risk. Likewise, the dominant sporting image of the brain-as-muscle is also making a comeback. As early as 1912, J. R. Hamilton argued that "the training of a brain is as easy today as the training of a muscle was yesterday. And no man with a well trained brain ever had to lie to his stomach about the dinner hour" (1912, 22). Hamilton directed his advice at prospective college students, but in more recent decades, potential brain trainees include a wide range of individuals, from businessmen to athletes. We continue to envision our brains as metaphorically muscular, as capable of change through training, and sometimes the physicality of controlling the brain dominates the description. In an EEGInfo video, Sue and Siegfried Othmer, parents turned practitioners, describe neurofeedback as a corporeal, almost bullying, process: "We're finding these ways of twisting the arm of the brain, if you will, and getting it to do our bidding" ("What Is Neurofeedback?" 2007).[8] Here, the brain is envisioned as requiring a bit of coaxing to fall in line and "do our bidding." Exercise training and accelerated-learning paradigms both advance this notion of "twisting the arm of the brain" through neurofeedback.

The traditional gym has not been replaced, but it has been supplemented. Of late, the emergence of EEGym, the Mental Gym, the Mindroom and other neurofeedback spaces have changed the landscape and definition of physical training. Elite athletes can expect to train not only their muscles (and their minds through sports psychology), but also their brains. Reaction times, peripheral vision, focus, relaxation, stress reduction under pressure—these are just some of the key areas that have been quantified by neurofeedback.[9] In this section, I focus on ways that conceptions of exercise have expanded to include aspects of brain physiology, specifically the ability to produce certain cerebral electrical activity on command. Current discourses of exercise inform our training of the brain and help to establish a new, ideal cerebral athletic physiology. Mindroom, NeuroTracker (a company that provides some of the equipment for the Mindroom), and Axon Sports are three outfits that exemplify these discourses. Not all of these companies and spaces are invested in EEG neurofeedback training, but they do engage in some form of technology-driven bio- and/or neurofeedback; more importantly, they engage and apply discourses of exercise to brain training and neurofeedback.[10]

Maintaining the Metaphor

Neurofeedback gymnasiums do not look or feel like traditional gyms, but they do share some of the key elements and assumptions. Take, for example, the Mindroom, which was responsible for the EEG neurofeedback training of the AC Milan soccer team. Contrary to popular media representations, the Mindroom is not a single place; in fact, there are several Mindrooms around the world. It would be better described as a style or approach to training that integrates psychological skills training (PST) with bio- and neurofeedback to help an "athlete achieve measurable and remarkable improvements in confidence and peak performance" (Mindroom 2015a).[11] One of the Mindrooms, run by Pierre Beauchamp, served the 2007–2010 Canadian short track speed skating team and served as the basis for several scholarly papers (Beauchamp, Harvey, and Beauchamp 2012). The actual spaces of the AC Milan Mindroom include a Sports Science Control Room and an Athlete Training Lab. Little information about the actual facility has been released to the media, but from the website (and video presentations), the Mindroom looks less like a gymnasium and more like a chemotherapy suite or a jack-in port from *The Matrix* (1999): a room filled with reclining chairs, each paired with playback screens and bio- and/or neurofeedback devices. In these images, it becomes clear that the focus of this athletic training is not below the neck but above it. Another sci-fi sounding company, Thought Technology, provides the Mindroom with its primary hardware and software suites.[12] Ironically, the Mindroom (dubbed the Mind Gym by athletes who use the facilities) is not termed the Brainroom— and this may be for good reason. One could speculate that a reference to the "mind" retains the standing of sports psychology and reminds the athletes that the work they do in the Mindroom is intended to strengthen their mental game, a game that has now been quantified through brain physiology. How this metaphoric muscle is exercised and trained depends on the discourses of traditional exercise that inform neurofeedback.

One of the most telling aspects of athletic neurofeedback training is the persistence of the gymnasium metaphor. Take, for examples, the names adopted by neurofeedback companies, including the MindGym, Mental Gym, and EEGym. On the one hand, it makes perfect sense: a ready-made metaphor that reinforces brain-as-muscle and inscribes certain rituals. As Millington notes, "It is not uncommon for a metaphor of the active body to be invoked in marketing texts to make brain training appear worthwhile" (2014, 497). On the other hand, applying the gymnasium metaphor to neurofeedback defies traditional definitions of training, which typically include mental components only as an *intangible* aspect of motivation and skill. Yet within the mental gymnasium those immaterial aspects of training are materialized and quantified—sometimes in problematic ways.

The gymnasium metaphor reinforces the idea that isolation and repetition are the best means of achieving what was once an unpredictable feeling of being "in the zone." Skills and movements are broken down, just as the set-up of a gym allows—and typically requires—clients to work particular muscles using specific equipment. In the case of the mental gymnasium, the (new) skill sets or focal points of training tend to be things like recovery, focus, and performance under pressure. These skills or abilities are being targeted in specific ways. For example, CogniScence Athletics, based in Montreal and the designer of the NeuroTracker training technique, recommends that "for most athletes two 8 min sessions a week provide a solid basis for rapid improvement in perceptual cognitive ability. Injured athletes may engage in more intensive training so that they can recover from the physical injury while actually heightening their mental game shape" ("NeuroTracker LowDown" 2011). NeuroTracker, which has launched a 3-D platform for athlete training, is one of the complementary training options available in the Mindroom. Its introductory videos explain that athletes can begin sitting down and advance to standing before moving on to actually performing activities within the chamber while still participating in mental training. "Begin your training sitting down, to isolate the brain; when you reach a high level, challenge yourself by standing up. After that, the only thing that limits your training is what you can imagine" ("NeuroTracker Sport Science Innovations" 2015).

Notably, early animal neurofeedback experiments depended on a similar logic of strength training: Barry Sternman trained cats to produce a sensorimotor rhythm (SMR) as a means of relaxation. The cats that were able to produce this rhythm were also more resistant to the toxicity of rocket fuel in a related experiment. Jim Robbins adopts metaphors of muscular training to describe these experiments: "What Sternman had done by teaching the cats to produce SMR, he would come to realize, was to strengthen their brain function at the sensory motor strip, the same way a person builds muscle mass by repeatedly lifting weights" (2008, 42). Robbins's characterization of Sternman's work is firmly situated in the current milieu of exercising the brain.

Alongside suggesting material benefits for the brain, the repeated exercises are a simple quantification of a potentially complex—and once intangible—state of mind. Through various routines, elite athletes are trying to increase specific electrical activity in their brains while quieting other activity in order to move willfully in and out of being "in the zone." In terms of EEG, these changes are quantified by the Mindroom as "improvement in Alpha, alpha 1, beta 2 and theta" (Mindroom 2015b). Yet the Mindroom's focus on peak sports performance (PSP), which brings the larger influence of sports psychology to bear on training regimes, remains vague. Mindroom lists the following attributes as characteristics of peak performance:

a. Clearly focused attention
b. Positive mental attitude
c. Physical and mental readiness
d. Positive competitive effect
e. Effortless performance
f. Perception of time slowing
g. Feeling of supreme confidence
h. Immersion in the present moment
i. A sense of fun and enjoyment (Mindroom 2015d)

A perfect cocktail of these characteristics results in athletes feeling "in the groove" or "in the zone," phrases "generally used in sports by athletes and coaches to describe a performance feeling where almost everything went right for them that day. However, athletes and coaches are often at a loss to explain how this state [is achieved] and/or repeat it" (Mindroom 2015d). When athletes are asked about their process, their answers sometimes equate physiological sensation with intangible states of being "in the zone." One member of the Canadian speed skating team, who started the bio- and neurofeedback routines after an injury, explains it this way: "I had to work on something else than only the physical part (because I was in a cast), so we started the mind part. I do it all together. I do the breathing so you don't think about anything, try to feel like I'm warm . . . no emotion now on the ice; that's what we're training to do" (Mindroom 2015b). Here, the primary measures of success for the athlete are warmth and an absence of emotion. The specific exercises and multiple repetitions that this athlete has undertaken have left him with some hacks and strategies for achieving what he feels is his optimal brain state. Is this "the zone"? Perhaps for him, it is; in any case, the gymnasium metaphor, which includes repetition and ritual, is alive and well in neurofeedback.

"Train above the Neck": Redefining the Athletic Body

At Axon Sports, a company invested in developing "cognitive training products," researchers and marketers have developed a new motto: "train above the neck." While the company does not rely on EEG, the basic sentiments expressed in its marketing materials help to illuminate the ways that the gymnasium metaphor informs a redefinition of the athletic body. A promotional video opens by discussing Malcolm Gladwell's ten-thousand-hour rule from his popular 2008 book *Outliers* (although they do not reference the original study: Ericsson, Krampe, and Tesch-Römer 1993), noting that ten thousand hours may max out an athlete—it may even be all the hours available before the body begins to deteriorate physically. This sentiment is echoed by David Linden, psychiatrist and professor of translational neuroscience at Cardiff University. In his book *Brain Control*, Linden argues that "too much practice (certainly in music and sport but probably also in surgery) can have the opposite effect [compared to enhancement] and even result in disabling conditions, such as cramps in the practicing limb. Mental training strategies are therefore a standard component of elite performance training in music and sports, and enhancing it with direct brain modulation would be a fascinating prospect" (2014, 106). In place of traditional physical training for ten thousand hours, Axon Sports suggests training a different muscle, "the athletic brain." When one "trains above the neck," one is working to improve certain aspects of the athletic brain: "high-speed decision making, visualization, emotional regulation, focus, reaction, and spatial reasoning" (Alessi 2012). And Axon Sports is not an anomaly.

CogniScience's NeuroTracker, is also marketed by touting the relative benefits of brain training as opposed to traditional physical training. It is "a unique training technique because it isolates areas of mental muscle that are critical for sport and improves them rapidly; this is because the brain structurally rewires itself if stimulated intensively and repeatedly. In the same way that muscle cells improve with physical conditioning, precise mental workouts improve the efficiency of neural networks. The main difference is that neural conditioning happens much faster and has a longer effect; this means that perceptual training with NeuroTracker brings increases in performance abilities at a superior rate to physical conditioning" ("NeuroTracker LowDown" 2011). The passage offers the gymnasium metaphors concerning training, conditioning, and regular workouts as well as an argument for training above the neck in addition to training the rest of the physical body: the relative speed of uptake and the lasting effects of the training. Paying attention to an athlete's physiological "load" via bio- and neurofeedback is also at the heart of the Mindroom. Bruno Demichelis (creator of the original Mindroom), stresses "the importance of physiology, which he said is often overlooked in the modern game. How much load a player has, then how he responds to it. Then, they analyze how a player performs during a game—the ultimate measure of a player" (Morgan 2015). Neurofeedback makes a player's physiological load visible, and therefore trainable.

The imperative to "train above the neck" implies that (1) to this point, an athlete's mental preparation has not been effectively executed, that training above the neck is something new and different than engaging the mental aspect of sports; (2) the brain is a new muscle situated in a liminal position: it can be trained just like the body *even as it is separate from traditional physical training*. Take, for example, a promotional video from Axon Sports in which the researchers being interviewed explain the placement of their cognitive training spaces in relation to the traditional gym: "We've set up what we call a lab and where we set it was actually adjoining to the training floor; the reason is that we want to communicate to the athlete that training above the neck is really a component of training. It's something you do in conjunction with what you're doing to train your body" (Axon Sports 2015). This explanation implies that the brain is separable from the body, which is being worked on in a different way. Whatever mental game preparation was taking place in the gym was not enough.

Finally, one might look to the ways that virtual sporting environments are at play in training above the neck. As the makers of NeuroTracker explain, the technology they employ "uses a true 3-D environment to activate the same neural processes that sports naturally activate; it then pushes athletes to their limits in a controlled scientific setting" ("NeuroTracker LowDown" 2011). Here, the worlds created in and for the brain are compared to the spaces and experiences an athlete might encounter in the sporting world. There is a sense that athletes can train and improve simply by activating the virtual—or better yet, neural—areas of the brain associated with playing and training for sports. "When you look at the future, the future is to marry this very sports specific cognitive training of the athletic brain with this very precisely tuned body and you have an athlete that redefines sport and performance as we know it today" (Axon Sports 2015). In the case of Axon Sports and many other mental gymnasia, the athletic body is being redefined by training above the neck. Yes, the athlete is expected to retain things like strength, flexibility, and speed, but the best athletic bodies will also have a brain capable of high-speed processing, regulation, and focus.

Rhetorics of Mastery: Advanced Brain Training

Even as the Mindroom applies discourses of exercise to brain training, other institutes and research centers offer new paradigms for thinking about learning and mastery. Advanced Brain Monitoring, a company run by Chris Berka, has developed a diverse portfolio of EEG technologies and software designed to intervene in aspects of learning, sleep, medical diagnosis, and treatment. These efforts demonstrate how paradigms of accelerated learning are redefining the concept of mastery, applying it not only to a task but to the conscious control of one's brain. One of the ironies in the new definition is that mastery of a task no longer means having the ability to "acquire complex skills and physiological adaptations" (Ericsson and Charness 1994, 725); indeed, mastery is only about equaling one's trainer. There is no way to move beyond or display an elasticity of understanding.

One of the oft-cited anecdotes about mastery is Ericsson et al.'s ten-thousand-hour rule: that is to say, if one seeks to be an elite performer in, say, golf or chess one would need to practice deliberatively for at least ten thousand hours (Ericsson, Krampe, and Tesch-Römer 1993). Ericsson's paradigm has been challenged in the decades since its publication, but in the common lore the ten-thousand-hour rule is still widely accepted. It relies on several assumptions: that learning can be quantified; that there is a specific level, "mastery," that can be achieved; that mastery is recognizable; and that only certain kinds of practice are valuable en route to mastery. Ericsson's paradigm has helped to define and delineate the speed and objectivity of learning in different ways. Numerous studies came before Ericsson—indeed, he and his co-authors detail this history within the paper itself—but his study established something in American culture such that new theories of learning often identify themselves with or against the ten-thousand-hour paradigm.

Of late, scientific and popular studies have challenged this long-view logic of mastery through various incarnations of "accelerated learning." Josh Kaufman, author of *The First 20 Hours: How to Learn Anything . . . Fast!* (2013), argues that through his five-step plan, individuals can learn new tasks and skills within the first twenty hours of deliberate practice. His book, which breaks down each step, urges learners to (1) identify a specific and tangible goal (not "learn to speak French" but, instead, "learn to speak enough French to order dinner at a restaurant"); (2) determine the specific sub-skills you need to practice to accomplish your goal (don't just read everything); (3) do just enough research about each sub-skill to begin executing it (don't wait until you've read everything); (4) eliminate environmental barriers to learning (keep your guitar out on a stand, rather than in its case in the back of the closet); and (5) be ready to commit twenty hours up front (ask yourself if this is really something you think you will enjoy) ("Accelerated Learning" 2013). Theories such as Kaufman's do not necessarily unseat the ten-thousand-hour paradigm, given that they are aiming not at mastery but at practice and skill acquisition; however, accelerated learning tends to set itself against Ericsson's paradigm.

And Kaufman is not alone; the market is flooded with books and programs promising to make learning easier and faster. Even the US military's Defense Advanced Research Projects Agency (DARPA) has invested in the accelerated learning paradigm, funding programs that ran from 2007 to 2012. Advanced Brain Monitoring (ABM) received one such DARPA grant; through it the company ran a series of EEG neurofeedback experiments designed to test whether it would be possible to exponentially advance the learning curve for novices in activities such as marksmanship and golf. In this section, I focus on three aspects of ABM's discursive tactics for changing how we understand mastery: (1) Reductively redefining mastery as a small set of skills necessary for a task by reading experts' brain activity, creating charts that illustrate when they are in the zone, and then teaching novices to match these patterns; (2) Defining mastery in terms of instrumental intimacy by changing the body's access to its own internal states via haptic sensors; (3) Characterizing mastery of a task through definitions of self-mastery and neuroscientific control. Berka's experiments raise questions about valuing and mapping expertise as a specific and desirable state, and they define learning in a particular way that may or may not indicate mastery.

Measure, Map, Match: Remastering Internal Brain States

Many accelerated learning training programs begin with elite athletes, the assumption being that if researchers can identify patterns of activity that recur in experienced athletes, then they will have identified the key elements of expertise. From heart rate to respiration to EEG activation, experts have mapped the bodies of elite athletes in search of this pattern. NeuroTracker, one of the technologies used by ABM, "was developed by testing Olympic athletes in Dr. Faubert's advanced lab. NeuroTracker data from a wide range of elite athletes is analyzed to extend the boundaries of sport science knowledge" ("NeuroTracker LowDown" 2011). Likewise, one of ABM director Chris Berka and colleagues' first experimental tasks was to illustrate that expertise *could* be mapped: "These studies indicate that psychophysiological profiles of expertise can be defined using EEG, though trends (in EEG alpha power, for example) may vary depending on the specific EEG frequency and hemispheric localization studied" (Berka et al. 2010, 4). By focusing on the elite athletes—the masters—Berka and colleagues have defined mastery as a fixed, quantifiable state to be achieved rather than the outcome of a process of learning, trial and error, and experience. Regardless of whether one definition or approach is better than the other, certain kinds of learning opportunities are foreclosed when one begins with the master and not the student.

Once they've completed the mapping, the research team can then devise various training paradigms that attempt to bring novices in line with the prescribed physiological blueprint, which has been termed the pre-shot peak performance (PSPP) state. For example, "marksmen consistently showed an increase in alpha power (seen in all electrode sites) beginning approximately two seconds before trigger pull, and a modest increase in theta power (seen only in right hemisphere frontal and central sites F_4 and C_4) beginning two-to-three seconds before trigger pull" (Berka et al. 2010, 7). By mapping the expert's electrical activation in the PSPP phase, the researchers are assuming that achieving the perfect shot is directly linked to and can be accomplished by matching this EEG signal. Theoretically, anyone from novice to expert should be able to accomplish this same result, if given enough feedback along the way.

Implicitly (if not explicitly), this type of paradigm values learning by rote rather than by experience: or as the old adage goes, they are giving students a fish rather than teaching them how to fish. Mastery means matching a pre-scribed blueprint for success. Progression to this point is accomplished largely through the use of haptic sensors that buzz or provide some other feedback when the novice reaches the desired state. "Three primary real-time feedback mo-dalities were provided: 1) audio feedback via headphones, 2) haptic feedback via two small (0.8cm diameter) shaft-less vibration motors attached behind the neck and 3) visual display on a projected screen and the remote computer" (Berka et al. 2010, 10). The desired states can vary depending on the task; primarily, as in the Mindroom, the optimal state has been defined as increased alpha and theta activation. This goal is true for each of Berka's experiments, which have included marksmanship, archery, and golf.

Instrumental Intimacy Revisited

From these experiments, Berka and her colleagues have created two sets of tech-nologies: the Adaptive Peak Performance Trainer (APPT) (which has desktop and real-time components), and the Interactive Neuro-Educational Technolo-gies (I-NET). The APPT is a "feedback mechanism designed to guide the trainee into the expert PSPP state" (Berka et al. 2010, 9); I-NET is a larger virtual environment suite that incorporates APPT. It is one thing to read about these studies and these technologies; it is quite another to see them in action, which is just what several journalists have done in the past few years. Morgan Freeman, host of Discovery Science's *Into the Wormhole*, and Adam Shaw of *BBC Horizons* each visited Chris Berka's lab to hear more about her research and the ways that current brain research has the potential to change the world. In both programs, novice archers and golfers are shown attempting to shoot a compound bow or swing a golf club while campy music plays over their awkward attempts. Then the neurofeedback technologies are introduced, briefly explained, and the novices try again—this time to amazing success. The haptic sensors tell them when they are "in the zone," which means that their electrical brain activity is within the peak performance range (defined in this case in terms of alpha and theta waves).

The instant success to which we are treated via the magic of television editing is not exactly accurate; however, the outcomes of Berka's experiments appear promising: "Novices in the APPT condition improved an average of 28.60% over the course of one training session. Novices in the Control condition im-proved an average of 12.22%" (Berka et al. 2010, 12). Or, as it is represented in *Through the Wormhole*, the "Adaptive Peak Performance Trainer (APPT) has shown to accelerate learning among novice athletes by 230 percent and audiences will see its application as archers conduct a training session guided by its intelligent neurofeedback" ("Discovery Science's" 2013). Here, a 20 percent improvement is converted into 230 percent acceleration.

The reports share the notion that conscious control of the brain depends on instrumental intimacy: in this case, external, sensory (haptic) signals. We do not know—or control—ourselves well enough as novices to modify our brain's electrical activity without some assistance. APPT appears to accelerate learning, but it should also remind us of the instrumental intimacies of chapter 1, in which EEG sensory equipment aided in the revelation and externalization of brain states. In those cases, the same electrical signals were translated into states of mood and arousal for all to see. The goal was not training, per se, but revelation and recognition of one's own—otherwise *inaccessible*—internal states. Here, the same electrical activation is also externalized, but in the service of creating mastery: mastery of a particular skill, and mastery of one's self.

Mastering One's Own Brain

Mastery of a task is one thing, but Berka's experiments also remind us that a large part of brain training, neurofeedback, and accelerated learning paradigms is mastery of oneself. In a recent television interview, Chris Berka explained that "what we're trying to do is give you the ability to control your mind and your mental state" ("Discovery Science's" 2013). Her statement involves at least three assumptions: (1) until this point, we did not have complete control of our minds; (2) mental states are connected to (and can be mapped from) brain activation—including the electrical activation of the brain; and (3) controlling the mind is a direct result of mastering the brain. Berka's remarks return us to the idea that neuroscientific control has been redefined by experts based on data collected from EEG wearables. Now that we have turned our gaze to the brain and devised various metrics for its activities, we have also refined the categories of mastery and conscious control. For Berka et al. the "goal is for you to train yourself to recognize that state and then, eventually, to move in and out of it at will" ("BBC Horizons" 2014). Mastery of a task is about the willful moving of the brain into and out of particular desirable states (as defined by experts).

Yet at the heart of this goal is a paradox: by improving one's ability to produce a particular type of electrical activity, one is also—according to the current neural learning paradigms—laying the groundwork for that activity, skill, task, or state of mind to become more automated. Like the metaphoric uncultivated field of weeds in need of a path, "repetition supports, strengthens and streamlines that pathway" ("What's Next" 2013). These are Chris Berka's words from her TEDx San Diego talk in 2013. She goes on to explain that "when you master a skill it can be automatically executed." Self-mastery, then, has been made into a conscious task that will eventually fall, once again, below our conscious radar. Therein lies the paradox of neurofeedback.

From Court to Corporation to Classroom— Biocybernauts All

The same companies and researchers involved in brain training and neurofeedback for sports have already begun to imagine applications outside of the gym. Possible additional applications include boardroom meetings, graduate school research teams, and individuals who suffer from various sleep disturbances (see chapter 3). Within this matrix, both in- and extra-lab applications have been suggested and developed.

One example is the Lotus, a small ceramic flower that communicates wirelessly with your smartphone and portable, dry, EEG headband to bloom when you are meditating correctly (i.e., producing a certain type of electrical activity in your brain). It is a low-budget neurofeedback device that helps to bring your brain wave activity in line with a standardized metric for relaxation. "'The number one thing, especially when starting [to meditate], is there's no feedback,' Rohan Dixit told *VentureBeat*. A neuroscientist who has measured the brain-waves of monks, he is one of three founders of the budding company. 'There's [usually] no confirmation that what you're doing is correct' when you meditate, he noted. Dixit points in particular to the audio feedback from the meditation teacher in the Lotus setup as being a key feedback aid" (Levine 2014). As we have seen throughout this chapter, providing feedback about brain wave activity through haptic processes allows users to effectively "feel" what they cannot see or (initially) sense and then make adjustments to bring themselves in line—if not with an expert then with a standardized metric.

In descriptions of the Lotus, echoes of the gymnasium metaphor persist: this device is about routine, repetition, and isolation of particular brain "muscles." "Just like having a gym buddy, meditating with friends will encourage you to stick to your routine, so we've baked social connections into the Lotus" ("Lotus" 2014). Unlike the Biocybernaut Institute, which can run into the tens of thousands of dollars for a week-long session, the Lotus costs about $150; and, although the Kickstarter campaign was not fully funded, the prototype (once called Puzzlebox or Puzzlebox Mind and now called Puzzlebox Bloom) is available for purchase (NeuroSky 2016). It is paired, perfectly enough, with one of NeuroSky's EEG devices (see chapter 1).

Neurofeedback and accelerated learning will continue to change the face of athletics and other dimensions of academic, corporate, and private life. More importantly, the discourses that inform such products and research are already in broad circulation: ideas of conscious control, the availability of newly quantified aspects of our physiology, and the development of new metrics for success and learning. As I demonstrate in the next chapter, these discourses extend even into sleep, which we have come to understand as an active state of mind ripe for intervention.

3

"Such a Natural Thing"

EEG Sleep Science in the Laboratory and Bedroom

> Sleeping seems to be such a natural thing that most persons
> believe they are doing a pretty good job of it. Since they are
> unconscious for eight, nine, or even ten hours, they naturally
> suppose they are getting plenty of sleep. But modern science has
> an unpleasant way of showing what a poor job we are really doing
> at many things which we think could not be improved.
> —DONALD LAIRD, *HOW TO SLEEP BETTER*, 1937

Sleeping has long been considered a "natural" and necessary part of most mammalian lives. We assume that if we are sleeping at all, we are successful. But are we? Since the early twentieth century, much sleep science research has focused on sleep disorders, those problems of insomnia, REM disruption, and circadian rhythm, among many others that keep so many of us awake during the wee hours of the night. Typically, the fear is that we are not sleeping enough, that jet-set business models are upsetting our natural rhythms, and that iPads, TVs, and smartphones are disrupting our endocrine systems. But these scientific inquiries are overwritten with cultural assumptions that vary by time and place. So we go in search of ideal sleep: in historical perspective and even in preindustrial societies. Failing this, we turn to the animal world and sleep labs for information about *why* we bother to sleep at all. Or we look to experts and wakeful heroes who seem to require little to no sleep. And after all of this searching and wondering, it might occur to us that sleep is certainly not so natural a phenomenon after all.

Our conception of sleep as an active—and malleable—state of mind is a relatively recent phenomenon; modern fears of lost sleep or problematic sleep changed in concert with the advent of mechanistic EEG sleep recording. While EEG has been associated with sleep science research since the 1930s, it enjoyed a heyday in the sleep laboratories of the mid-twentieth century. Discourses of monitored, medicalized, and measured sleep emerged after Eugene Aserinsky and Nathaniel Kleitman recorded and correlated EEG brain activity with eye movement in the 1950s. We now term their observed phenomenon "REM sleep," but at the time Aserinsky and Kleitman were working against the grain of accepted theory and practice. Sleep was considered to be, fundamentally, a state of mind in which nothing happened; Aserinsky and Kleitman turned sleep research on its head by suggesting that sleeping was a physiologically active state of mind, one that could be recorded, measured—and even, perhaps, improved on. Contemporary EEG wearable technologies, such as Kokoon and Neuroon, rely on discourses of active sleep (and sleep problems such as jet lag and shift work) that emerged in the scientific experimentation and print news media of the 1950s. And these recent technological innovations have lured EEG out of the laboratory and into the bedroom—or wherever we find ourselves in need of "better" sleep.

In this chapter, I first complicate assumptions about the nature of sleep via the work of scientists and sleep historians; next, I parse several decades of sleep science research to focus on the advent of brain wave discourses that accompanied the advent of EEG-REM monitoring and therapy in the scientific literature and print news media of the 1950s and 1960s. Finally, I analyze two of the latest EEG technologies to enter the bedroom, Kokoon (headphones) and Neuroon (eye mask), which ostensibly offer us access to and further regulation of this seemingly unconscious brain state that has been slowly rendered accessible, malleable, and controllable. Along the way, I unpack the intersecting discourses of nature-culture, control, and optimization.

The "Nature" of Sleep

It remains an astonishing anachronism in the history of science that Watson and Crick unraveled the structure of DNA before virtually anything was known about the physiological condition in which people spend one-third of their lives.

—CHIP BROWN, "THE STUBBORN SCIENTIST WHO
UNRAVELED A MYSTERY OF THE NIGHT"

Natural sleep has many definitions. Ask Jessa Gamble, science writer and author of *Siesta and the Midnight Sun: How Our Bodies Experience Time* (2011), and you are likely to hear stories of biological, chemical clocks that have been disrupted by modern technology. Ask the residents of Bad Kissenger, a Bavarian spa town in Germany, and you would likely hear about the importance of "chronotype," or "a person's preferred sleep pattern" (Beck 2014, 2). Ask Jerome Siegel, a professor of psychiatry and director of the UCLA Center for Sleep Research, and you are likely to hear references to our preindustrial ancestors. Collectively, we can say and have said a lot about sleep, but we know very little beyond the basics: we cannot survive without it (Rechtschaffen 1998); beyond that, "sleep is mysterious, and we don't totally understand why we need it—just that we do, and bad things happen if we don't get enough of it" (Beck 2014, 2). We take for granted that sleep should be a particular way, should last for a particular length, even that we should sleep at all. I delve into the sciences of sleep in the next section; first, I examine some of the ways that sleep has been represented in or as a "natural" state. In this section, I briefly contextualize contemporary sleep patterns and expectations in a longer genealogy of practices in order to argue that sleep is a necessary, and strangely naturalized state of being.

In an article published in *Current Biology*, Gandhi Yetish and colleagues pose the question, "How did humans sleep before the modern era?" (2015, 2862). Part of the problem with such a question, they suggest, is that "the tools to measure sleep under natural conditions were developed long after the invention of the electric devices suspected of delaying and reducing sleep." As an answer, the researchers devised an experiment to "investigate sleep in three preindustrial societies" (2862): the Hadza Hunter-Gatherers of Tanzania, the Tsimane of Bolivia, and the San people of Namibia. Participants wore an Actiwatch-2 for six to twenty-eight days and also wore monitors to measure their body temperature. While the results of the study advance certain aspects of sleep research (for example, that body temperature and ambient temperature play a greater role than ambient light in the sleep cycles of these three groups), the study's design is based on deeply problematic assumptions.[1] To be blunt, in order to approach "natural sleep," the authors compare preindustrial peoples with "pre-modern era *Homo sapiens*" and argue that their study is thereby able to "provide insights into the nature of human sleep under natural conditions" (2862–63). Indeed, in a press release about the study, project leader Jerome Siegel argues that "many of us may be suffering from the disruption of this *ancient* pattern" (Sullivan 2015, emphasis added).

What I find so interesting about this study is not only the disturbing conflation of less industrialized modern-day peoples with "our ancestors" in the popular news release (Sullivan 2015) but also the seemingly easy move toward arguments from and about nature. Siegel claims that "rather than saying modern culture has interfered with the natural sleep period, this is a case in which modern culture, with its electric light and temperature control, was able to restore the natural sleep period, which is a single period in traditional humans today and therefore likely in our evolutionary ancestors as well" (Sullivan 2015). This statement is mirrored in the *Current Biology* article: "Our findings indicate that sleep in industrial societies has not been reduced below a level that is normal for most of our species' evolutionary history" (Yetish et al. 2015, 2867). The implication being that the "natural" is accessible to us via data collected from contemporary, preindustrialized participants. When such arguments are followed to their flawed conclusions, as they are in another news article about the study, journalists can use the experiment as a basis for reassuring readers about their nightly averages—"scientists say the industrial world people can finally relax about sleeping less than eight hours a night"—and, quoting Siegel, conclude that "we find that humans living in a natural environment do not sleep more than modern-day humans" ("8-Hour Sleep Myth Debunked" 2015). As one might suspect, Yetish et al.'s study is not the first of its type, or even an outlier; during the boom in periodicity studies in the 1990s, shutting men and women into light-controlled laboratory spaces resulted in science papers on biological clocks and *New York Times* articles with references to ancestral humans, such as "Modern Life Suppresses an Ancient Body Rhythm" (Angier 1995). Yetish's paper is in part a response to such studies—and to other research in the history of science.

Several histories of sleep illustrate that the "natural" phenomena of sleep would be better described as a varied and naturalized experience. In 2012, journalists reveled in revealing the eight-hour sleep cycle to be a "myth" (Hegarty 2012). The report was based largely on books by historians Roger Ekrich and Craig Koslofsky. News reports were quick to latch onto one of Ekrich's final points, that there is a long and varied historical record of bifurcated sleep in Europe. However, his book, *At Day's Close: Night in Times Past*, presents a much more complex portrait of sleep, relying on three decades of archival research from which Ekrich concludes that, yes, several European classes often experienced segmented sleep—or a period of wakefulness bifurcating the hours of darkness. However, Ekrich does not argue that all people historically spent the night in bifurcated sleep; instead, he notes that, even historically, sleep depended on one's class position, labor, environment, health, and many other factors. Likewise, Koslofsky, in *Evening's Empire*, argues that nighttime was equally complex in the Early Modern period; he notes that "sleep is the first necessity of the night" (2011, 6), but alongside this activity, he also includes work and leisure. These studies do not demonstrate that there was once a natural state of sleep but, instead, that sleep has changed over time, adapting to various environments, lifestyles, and necessities.

Bringing the history of sleep into the modern era, Alan Derickson makes a cogent argument for the linkage of postindustrial work and sleep practices. His book, *Dangerously Sleepy: Overworked Americans and the Cult of Manly Wakefulness*, brings to light many of the sleep discourses that have led to associations between sleep, weakness, effeminacy, and laziness. With the exception of a brief respite in the postwar 1950s (on which, see more below), Derickson argues that American men have been raised on a diet of sleeplessness: "for more than two centuries, a chorus of influential voices, virtually all male, has proclaimed sleep a vice and sleep deprivation a virtue" (2013, 1). From Benjamin Franklin to Thomas Edison to Donald Trump, Derickson charts the disciplining of (American male) physiology away from sleep and toward omniscient wakefulness. He also tackles issues of shift work and labor laws that have not rendered sleep a top priority for American corporate interests (issues directly relevant to the development of the latest EEG sleep technologies, Kokoon and Neuroon).

From the advent of alarm clocks to lost agrarian and seasonal indicators, the modern era has changed expectations about and desires for sleep. Far from being a "natural thing," sleep varies by time, place, and culture. Consider the small German spa town of Bad Kissenger, a tourist-driven medicinal retreat. The town's leadership decided to update its amenities to include accommodations for residents' and visitors' individualized sleep patterns—the argument being that humans have distinct chronotypes and that by living within these partially genetic, partially environmental strictures, one can live a more productive and healthy life. In this case, finding the natural means aggregating individual data and making structural adjustments, as "most people are socialized not to sleep when their bodies naturally want to" (Beck 2014, 1). Bad Kissenger's leadership seeks to resocialize its residents and visitors—to help them understand their natural bodily needs. Working in conjunction with Thomas Kantermann (a chronobiologist), the town's business director, Michael Wieden, seeks to make Bad Kissenger "a place where your internal time is acknowledged" (9). As this example makes explicit, idea(l)s of natural sleep have, and will continue to take on, myriad forms. There is no "natural" to rely on, or fall back to, when it comes to sleep. Instead, we have scientific and historical inquiries into a phenomenon that is largely defined and framed by political, social, economic, and scientific contexts.

REM and EEG: Sleep Science Evolved

Two score and ten years ago Aserinsky and Kleitman brought forth on this planet a new discipline, conceived at night and dedicated to the proposition that sleep is equal to wak[ing].

—WILLIAM DEMENT, ADDRESS TO THE ASSOCIATED

PROFESSIONAL SLEEP SOCIETY, 2003

As Jerome Segal's study of "natural" sleep indicates, what we know about sleep has long depended on the technologies we employ for its measurement and evaluation. Mechanized sleep research is largely a mid- to late-twentieth-century phenomenon. Early-twentieth-century research often depended on a sleeper's movement (measured by a somnokinetograph), heart rate (measured through EKG), or even stomach contractions (Laird 1937, 63).[2] But some of the biggest breakthroughs in sleep science came with EEG: first, with Berger's recognition that there is a difference in the brain's electrical activity when one's eyes are open or closed (1929); then, with the recording of five sleep stages by Loomis, Harvey, and Hobbart (1937), followed by the first studies to introduce REM sleep as measurable via EEG by Eugene Aserinsky and Nathaniel Kleitman (1953); and finally, with Michel Jouvet's conception of sleep as an "active" state (1959).[3] In this section, I discuss the ways that EEG-enabled constructions of REM sleep changed the basic discourses about sleep's form and function between 1950 and 1970.

As we saw in chapters 1 and 2, instrumental intimacy becomes possible once there is a basis for intervention: a physiological phenomenon that has become the target of scientific measurement and therefore subject to potential modification. In the case of EEG and sleep, one of the earliest turning points was Aserinsky and Kleitman's study of the relationships between eye movement, EEG, and dreaming. Published in 1953 and picked up by the media in the late 1950s and 1960s, the study not only introduced the neurophysiological community to the concept of what would become known as rapid eye movement (REM)—now considered a distinct and *active* stage of sleep—but also linked eye movement to dreaming, noting that their "method furnishes the means of determining the incidence and duration of periods of dreaming" (1953, 274).[4] Aserinsky and Kleitman's experimental write-up is quite modest—two and a half columns in *Science*, but it contains this consequential statement: "The fact that these eye movements, EEG pattern, and autonomic nervous system activity are significantly related and do not occur randomly suggests that these physiological phenomena, and probably dreaming, are likely all manifestations of a particular level of cortical activity which is encountered normally during sleep" (274). Put another way, Aserinsky and Kleitman not only recorded what we now call REM sleep but also were the first to comprehensively record and associate specific eye movement and brain activity, fundamentally challenging then-contemporary assumptions about the inactivity of the sleeping brain.

Importantly, Aserinsky and Kleitman did not bring REM into the world fully formed; instead, the authors spoke of "rapid, jerky eye movements" (1953, 274). And as Claude Gottesmann and Adrian Morrison have argued, the naming of REM sleep was far from a simple matter. Proposed names continued to crop up in the decades after Aserinsky and Kleitman's study. Among them were "rapid sleep, Stage 1, Phase 1REM, rapid eye-movement sleep, activated sleep, active sleep, sleep low-voltage fast, dreaming sleep, emergent Stage 1, PRS (rhombencephalic phase of sleep), PS (paradoxical sleep), fast wave sleep, desynchronized sleep, deep sleep, light sleep, and para sleep. . . . [U]nderlying the quest for an appropriate name was the struggle to place it within the framework of the brain's workings as they were known then" (Morrison 2013, 401). The naming debate illustrates the paradigm shift necessary for the neurophysiological community to accept Aserinsky and Kleitman's results, which went beyond the naming of a new phase of sleep. Rapid eye movement alone was not the primary importance of Aserinsky and Kleitman's study; instead, it was their hypothesis that patterns of eye movement that coincided with EEG-measured brain activity meant that the brain was active during sleep.

Thus, what is most remarkable about Aserinsky and Kleitman's work is its break from both neurophysiological research, in which "sleep was easily equated with inactivity and rest" (Dement 1990, 18), and from psychoanalytic hypotheses about sleep, dreams, and accessibility of mind. Aserinsky and Kleitman's experiment paved the way for studies of the "particular level of cortical activity" which is a part of "normal sleep" (1953, 274). As Adrian Morrison argues, "Important figures in neurophysiology and psychiatry were forced to realize that REM had overturned the accepted view that sleep was a simple state of withdrawal from the outside world, quiescent in nature, and that a portion of sleep had the 'strange' characteristic of an EEG pattern that was essentially like that of wakefulness" (2013, 393). Morrison's article traces what he calls this "new" state of consciousness through its development at two of the major conferences of the era—the Ciba (1960) and Lyon (1963) symposia, which featured scientists such as Dement (Kleitman's student), Jouvet, and others—that shaped REM sleep research for decades to come.

David Foulkes, who became a major player in sleep research in the 1960s and 1970s, also explores the reshaping of the field of neurophysiology after Aserinsky and Kleitman went public with their findings. In "Sleep and Dreams, Dream Research 1953–1993" (1996), Foulkes, a rival scientific figure who wished to complicate associations between REM sleep and dreaming, helps to explain the explosion of funding and laboratory space dedicated to sleep research in the 1960s:

> In 1964–1965, NIMH had "supported over 60 projects related in whole or in part to studies of sleep and dreams, with awards totaling over $2,000,000" . . . (Because other federal agencies also were supporting the area, even this figure understates the government's overall commitment.) Relatively speaking, 1964–1965 must have been close to, if not at, the all-time highpoint at which any nation ever has supported basic dream and dream-related research (bearing in mind the more intimate relation of sleep research and dream research at that time). (1996, 613)

Here, we see the sheer force of funding applied to this new state of consciousness. For a field that began with a few scattered scientists, sleep science gained quite a bit of traction in the decades following Aserinsky and Kleitman's research.[5] "The recognition of REM forced a re-evaluation of the organization of consciousness among brain scientists at the beginning of the decade of the 1960s, even among some of its leaders. This new, 'paradoxical' state led to a tremendous outpouring of research during the first years of the decade, perhaps unprecedented in the world of science" (Morrison 2013, 405). The hyperbolic excitement of Morrison nearly seventy years later indicates the impact Aserinsky and Kleitman's study had on the neurophysiological world.

Paradoxically, the association of sleep with brain activity was not fully recognized or accepted by the field for nearly a decade. As Dement notes in an autobiographical retrospective of his own follow-up findings concerning EEG-based brain activation during sleep, gatekeeping helped the field retain a strong force of inertia. His own papers linking sleep and dreaming, which are based on presumptions of an active brain, were met with much skepticism and derision:

It is very difficult today [circa 1990] to understand and appreciate the exceedingly controversial nature of these findings. I wrote them up, but the paper was nearly impossible to publish because it was completely contradictory to the totally dominant neurophysiological theory of the time. The assertion by me that an activated EEG could be associated with unambiguous sleep was considered to be absurd. As it turned out, previous investigators had observed an activated EEG during sleep in cats but simply could not believe it, and ascribed it to arousing influences during sleep. . . . [One of my colleagues] was sufficiently skeptical that he preferred I publish the paper as sole author. After four or five rejections, to my everlasting gratitude, Editor-in-Chief Herbert Jasper finally accepted the paper without revision for publication in *Electroencephalography and Clinical Neurophysiology*. (Dement 1990, 23)

Dement's work (Dement and Kleitman 1957) was later confirmed in animal studies by Jouvet (1959) and Hubel (1960), and in humans by Goodenough et al. (1959); however, his initial experience at the hands of the neurophysicological field's gatekeepers as late as 1958 indicates the reluctance of some to recognize sleep as an active state of mind—or "new" state of consciousness.

From the vantage point of the early twenty-first century, it is difficult to imagine a time when sleep was not considered by scientists to be an active state of mind. Yet Aserinsky and Kleitman's work is only sixty years old, and its acceptance is far more recent. In the next section, I discuss some of the routes through which the discourses of sleep as active and malleable state entered the popular consciousness of mid- to late-twentieth-century America. Our sense of instrumental intimacy and its application to sleep begins in these decades and continues to inform mobile EEG technologies such as Kokoon and Neuroon.

Periodical Fascination: REM Sleep in the News, 1950–1970

There may be "something cooking during sleep that we don't know much about."   —*WASHINGTON POST*, 1960

In the print news media of the era, word of Aserinsky and Kleitman's study was slow to spread. A year-by-year Proquest search of national news stories turned up only three stories that touch on Kleitman's research and the use of EEG to study dreaming before 1960—only one of which actually focused on Kleitman's experiments and reports of REM sleep.[6] Despite their relative absence from the news until the 1960s, Aserinsky and Kleitman's experimental work is character-ized from the 1960s onward as a watershed for sleep research.[7] After Kleitman's research recategorized sleep as an active state—at least during rapid eye movement—the print news media helped bridge the gap between the theory and the potential for instrumental and human intervention.

A recurring, syndicated column, "How to Keep Well," by Dr. Theodore R. Van Dellen, exemplifies the pre-1960s print news media coverage of sleep. Reporting on sleep research among other medical phenomena, Van Dellen mentions EEG, but remains focused on the sleep stages named in the 1930s, concentrates on the depth of sleep, and makes no mention of REM sleep or its associations with meaningful brain activity into the 1960s. In one column from 1956, the good doctor asks, "How well do you sleep?" His answer—as we might suspect—begins by noting the measures applied to sleep: "Many tests have been devised, including the use of the electroencephalograph (EEG). This machine records the brain waves which are different during sleep than when awake. These waves follow a certain rhythm. The waking pattern is re-placed at night by altered wave forms and slower frequencies. In deepest sleep, the waves are of large amplitude, called delta waves." He goes on to describe the current testing scheme in which "the individual is put to bed and the EEG electrodes are fastened to the head. A two-minute record is made at five-minute intervals for seven hours. The tracing is read and the depth of sleep for each test period is graded. One point, for example, characterizes a dull drowsy stage; the presence of delta waves indicates full unconsciousness and is worth four points. Periods of wakefulness receive zero. A perfect score is 336 but actual per-formance is much below level." Note, here, that there are really only two points of interest for the scientists testing sleep: delta waves associated with "full un-consciousness," or deep sleep, and wave forms associated with "periods of wakefulness." Despite Kleitman's research findings, periods of lighter sleep are not yet associated with REM, sleep quality, or even the act of dreaming. This absence of REM/dreaming as possible points of examination is con-firmed by Van Dellen's summary: "During the remainder of the night [after the first period of deep sleep] the brain waves change in a cyclical manner varying from a pattern similar to that when awake to typical delta waves of deepest slumber" (1956, 56). For Van Dellen, the periods of time between deep sleep and wakefulness are uninteresting and unworthy of note. Indeed, it is not until 1960 that the column reports on eye movement, dreaming, and more specific measurements offered by instruments such as EEG (Van Dellen 1960, 20).

The year 1960 is a watershed moment for Aserinsky and Kleitman. Nearly all of the national news reports about sleep between 1960 and 1969 make mention of Kleitman (his graduate student Aserinsky is occasionally forgotten) and concern potential linkages between EEG and REM sleep. In part, this is due to the publication of several popular books about sleep and the surge in sleep research: science writer Edwin Diamond's *The Science of Dreams* ([1962] 1968); science writer Gay Gaer Luce and psychologist Julius Segal's *Sleep* (1966); and psychologist David Foulkes's *The Psychology of Sleep* (1966). In a review of the latter two books, a *New York Times* reviewer likens the recent sleep research to the space race: "In the past ten years there has been world-wide publicity for the American probes into space. Much less attention has so far been paid to another type of scientific exploration in which American researchers have made a breakthrough that may prove even more important to human health and happiness than the launching of astronauts—the systemic study of the mysterious processes of sleep" (MacKenzie 1966, 307). His next sentence lauds none other than Nathaniel Kleitman—a trend that marks the press coverage of sleep research throughout the 1960s.

A piece in the *New York Herald Tribune* covering Edwin Diamond's book *The Science of Dreams* characterizes Kleitman's research on EEG as germinal: "The discovery took dream research out of amateur psychological speculation and put it on a solid scientific basis" (Ubell 1962, F4). In Edwin Diamond's own column about dreams for the same newspaper, he argues, "In the past, before the new view of dreams made possible by the Kleitman method, speculation about dreams in sequence had been rare, principally because few investigators had any idea there was any sequence to speculate about" (1962b, 41). In an earlier piece, Diamond sets the stage for his own research, terming Kleitman a "hero" and instating him as "the world's leading authority on sleep":

As frequently happens in scientific research, the great passkey to the inner world of dreams was discovered quite by accident. The inadvertent hero was a nameless Chicago infant, asleep in his crib one day in 1952. The real hero was Prof. Nathaniel Kleitman, the distinguished University of Chicago physiologist. At the time, he and one of his graduate students, Eugene Aserinsky, were studying the sleep habits of infants during the first six months of life. Prof. Kleitman has been called, with justice, the world's leading authority on sleep. His research, among the most imaginative, yet meticulous, work in modern science, provided the essential physiological framework for understanding sleep and dreams. (1962a, 36)

Throughout the 1960s, other reporters continue this deference to Kleitman. Harry Nelson, who covered several stories about sleep research during this period for the *Los Angeles Times*, declares "research on sleep has been booming since 1953 when Dr. Nathaniel Kleitman, now in retirement in Santa Monica, discovered it is possible to tell whether a person is dreaming by watching the movement of his eyes" (1965, B1). We hear similar sentiments from John Davy in the *Observer*, who notes that "perhaps the central discovery of sleep research was made in the early 1950s when Nathaniel Kleitman and Eugene Aserinsky, at the University of Chicago, discovered that people woken in this [REM] phase of sleep almost always reported a vivid dream. This whole REM state has proved, on investigation, to be quite extraordinary" (1967, 13). Authors Luce and Segel argue in a *New York Times* article that "the association between sleep and dreaming was demonstrated in 1952 by Eugene Aserinsky and Nathaniel Kleitman at the University of Chicago, a finding that inspired worldwide research, for it offered a means of telling from outside when a person is dreaming" (Luce and Segal 1966a, 29). Or, as David Foulkes argues:

> That was a period, perhaps unprecedented elsewhere in psychology, in which age-old questions suddenly seemed both resolvable and resolved. The heady atmosphere of this brief period, captured by, among others, Diamond, Trillin, and Luce and Segal, was soon to give impetus to a vast expansion in empirical research on dreaming, to the formation of a specialty organization where this research was presented and discussed [the Association for the Psychophysiological Study of Sleep (APSS), which met first (but under no particular name) at the University of Chicago in 1961], and to the creation of a new multidisciplinary research specialty ("sleep and dream research"). (Foulkes 1996, 610)

Here, we have the first inklings that research into REM sleep and dreaming via EEG bode major changes in the understanding of and potential interventions into this newly minted state of mind.

The Instrumental Intimacy of Sleep

Arguably, Aserinsky and Kleitman's watershed report created an opportunity for sleep to become a more succinct product and producer of brain wave ideologies and instrumental intimacy. Embedded in and informing these re-ports are discourses about mechanistic access and control. Three tenets of instrumental intimacy relate directly to sleep: (1) The move away from intro-spection to machine-mediated seeing. In the case of sleep research, sleep and dreams are redefined as physiological states of mind that should be studied apart from psychoanalytic methods; (2) The belief that internal states of mind (includ-ing dreaming) are technical phenomenon, which include stages and recognized electrical patterns; as such, they can be accessed, externalized, and modified; (3) The expectation that instrumental data allows for optimization of various states of mind, including sleep. Individuals can adjust their sleep stages so as to better adapt to modern life, including shift work and jetlag.

First, thanks to EEG, scientific sleep research moved away from the realm of psychoanalytic speculation and theory and became more firmly entrenched in the realm of physiology. David Foulkes, for example, observes that "with the advent of the cognitive approach, there was a clear shift of focus from dreams, the product, to dreaming, the process" (1996, 619). Within the print news me-dia, there was an initial desire to reconcile psychoanalysis and physiology. A tag line from a *New York Times* report claimed that "scientific explorations of dream-land are confirming some psychoanalytic theories about the life we lead in sleep." In this extended piece, Leonard Wallace Robinson champions the re-search of Kleitman but sees sleep research, in general, as building on Freud's "educated" and "brilliant guess" (1959, 52) about our unconscious lives. Yet Robinson is an outlier. News reporters in the 1960s would, sometimes quite adamantly, characterize the EEG sleep research of the 1950s and 1960s as a break from psychoanalytic speculations. Take, for example, this early report on Kleitman's work:

It took a plain talking, scientifically minded physiology professor at the Univer-sity of Chicago to bring the dream world down to earth for some common sense analyzing. He is Dr. Nathaniel Kleitman, generally recognized as an authority on the physiological process of dreaming. . . . Poets may not like it, but Dr. Kleitman has unhitched many a dream from a star by hitching the dreamer's head to an electroencephalograph and cardiotachometer to study the brain waves, heart beats, body and eyeball movements during sleep and dreams. . . . They leave the crystal balls to the soothsayers and the Oedipus complex to the psychoanalysts. They are not interpreters of dreams. Their interest is in the physiological dream process, not the psychological content. (Browning 1956, 17)

In this telling, science trumps speculation, instrumentation tops poetic rhapsodizing, and physiological measures outdo psychoanalytic analyses. Quite literally, dreams have been situated in the brain and therefore, the body of the dreamer.

Included in this push to separate sleep research from psychoanalysis are Allan Rechtschaffen and Joe Kamiya's dream studies. Rechtschaffen was the researcher who first noted that lack of sleep was fatal in rats, and Kamiya, who was at the University of Chicago with Kleitman, also worked on neurofeedback (see chapter 2). One front page story in the *Washington Post* reports on their research with excitement: "Dream research groups, including two at the University of Chicago, are revealing fascinating new worlds in the life of the body during sleep. In Chicago, professors Allan Rechtschaffen and Joe Kamiya are probing with electronic devices an area that previously only Sigmund Freud, using guesswork, inference and insight, explored to reveal the symbolic nature of dreams. Rechtschaffen and Kamiya agreed that the 'new method,' called EEG, already has proved with physical data what psychoanalysts have only suspected." When asked why they pursue these studies, Kamiya replied, "We want to show that the dream is a bodily activity. . . . There is a real physical basis to the dream that is remembered. It is not merely a thought. It has long been felt that thoughts, ideas, hopes, fears, were not grounded in man's physical system. We want to get rid of this dualism in our thinking, that there is a realm of the mind and a realm of the body" ("Psychologists Get Some New Ideas" 1960).

The break from psychoanalysis authorizes a physiological basis for and scientific theories concerning a phenomenon once deemed to be ephemeral and inaccessible. Indeed, making dreams—and more importantly the sleep stage whence they emerge—into physiologically accessible states of mind is a move championed in a rare 1950s news article about Kleitman and his work. The scientist, himself, is quoted as stating, "No matter how you slice it, dreaming is a normal physiological process, just as normal sleep is, and as universal" (Browning 1956, 17). And this sentiment is also present later in Norman MacKenzie's review of recent popular books about sleep research: "We now know . . . that dreaming is a regular and biologically regulated characteristic of sleep, and that the state of dreaming is marked by so many distinct physiological features that we must assume it plays a vital (though as yet unknown) role in our lives. Hitherto, the best guesses about this role were psychological." Note the break from psychoanalysis that remains a necessary feature of such arguments.

Indeed, MacKenzie goes on to include specific comparisons between psychoanalysis and the advancements of modern physiology: "The new discoveries do not mean that we can now simply discard the immense amount of work that has been done on the psychological aspects of dreaming—work that was done, inevitably, under crippling limitations. But they have given us the means of setting the psychology of dreams in their proper physiological context, and thereby understanding better the complex links between mind and body" (1966, 307). Of note here are the "crippling limitations" of Freud and his peers and the assertion that through new research paradigms we are finally "setting the psychology of dreams in their proper physiological context."

A second discursive move, which follows from the first, is to transform sleep from a mysterious (dark) state that we fall into (or fail to fall into) to a state of mind that is illuminated, accessible, and impressible. In an article for the *Irish Times*, Michael Viney focuses on the strides sleep science has made beyond psychoanalysis. He zeros in on the work of William Dement (a student of Kleitman) and Ian Oswald, through which we can see that post-REM EEG seeks to render sleep as a physiological, rather than psychological, state, a move that makes sleep a physical phenomenon that can be quantified. "For the fact is, until the last two decades we knew remarkably little about the state—or states—in which we spend about a third of our lives. For most of physiological and psychological science, sleep was a relatively simple phenomenon of obvious function. By learning more of its bio-chemistry, its effects upon behavior, we are almost certain to come upon discoveries with explosive implications for the treatment of physical and mental illnesses." Before we had the ability to measure various aspects of sleep and chart its properties and vicissitudes via electrical impulses in the brain, Viney claims, "sleep was a relatively simple phenomenon of obvious function" (1966, 10).

Viney's description is seconded in several other publications of the era. In their extended report on sleep for the *New York Times*, Luce and Segal make a more explicit division between our previous state of ignorant bliss and our current state of technophilia:

Sleep study is not new, but the pace of current discovery is startling. A vast new library of information . . . has been compiled in the United States, Western Europe, Russia, Mexico, Japan, and elsewhere. It reflects the availability of computers and new scientific instruments . . . within a decade this work has revolutionized our concept of sleep. . . . Sleep may feel like a blanket of darkness, punctuated by dreams, a time when the mind slept. Nothing could be less true. All night a person drifts down and up through different levels of consciousness as on waves. With the EEG machines, which trace the pattern of the brain's electrical changes, and other sensors that record body temperature, pulse, respiration and the like, researchers have charted the stages of the average person's long night's journey into sleep. (1966b, 29)

Here, the mythic version of sleep as a "blanket of darkness" when "the mind slept" is countered by our ability to measure the brain's activity. In short, because we can gather more information about the brain's electrical activity, we can no longer claim that the mind is unavailable, unconscious, or—ironically—asleep. We hear a similar logic in Luce and Segal's description of sleep laboratories of the 1960s: "The preparations [in sleep laboratories]—centering around an amplifying and recording device known as an electroencephalogram, or EEG machine—suggest a pilot's instrument check before take-off and the keen sense of anticipation among sleep scientists—psychologists, psychiatrists, chemists, biologists, neurophysiologists and mathematicians—contrasts markedly with the incurious acceptance with which most of us submit to sleep" (29). The instrumental preparations listed and the number of scientific branches interested and involved in sleep research paints a tableau of technical intervention into a state of being that has long been taken for granted. Even—and especially—if sleepers protest that they do not have access to the realm of sleep or that they do not dream, the technology can illustrate what is really going on: "Many a sleeper who boasts that he never dreams has been proven wrong by the electronic devices that record brain wave patterns and eye movements" (Browning 1956, 17). And with this statement, we have reached a point in the discursive development in which the instruments know more than the sleeper.

The third tenet of instrumental intimacy as it relates to sleep is that sleep stages can be optimized for modern life. Although links between the specific discourses of shift work, jetlag, and optimization through EEG do not fully blossom until the turn of the twenty-first century, hints of them can be found in the print news media of the 1960s. As I discussed in chapter 2, the 1950s and 1960s saw a move from relegating the unconscious mind to psychoanalysis to modeling the once-unconscious as an instrumentally accessible consciousness—a state of being that can be accessed via new instrumentation and, as such, be accessed for the sake of intervention and change. As one reporter succinctly puts it, "The old dichotomy between conscious and unconscious thinking is now clearly too crude to be sustained" (MacKenzie 1966, 307). Luce and Segal state things even more bluntly: "As a byproduct of EEG laboratory studies, future man may come to know and to exercise control over many now unrecognized states of consciousness and their allied effects in the body" (1966a, 73). Reporters of the era—the same era as Kamiya's brain-training studies covered in chapter 2—expend much ink discussing the possibilities for control. Typically, they first establish that what was once deemed unconscious should be redefined; sleep research is no exception to this phenomenon:

> It is also clear that in some way or other we are conscious even in deep sleep. Mothers will sleep as a loud jet flies overhead, but wake at the smallest whimper of a baby. Suggestible subjects will respond to hypnotic commands in deep sleep. . . . But experiments with volunteer nurses in New York have shown that people can even learn to recognize their own sleep states while asleep. They were asked to sleep with a small switch taped to a finger, and to press it twice when dreaming and five times when asleep without dreams. One Philippine girl learned to distinguish between her own REM sleep and Stage II sleep. All of this suggests that we have the rudiments of unexpected powers over our sleep states. (Davy 1967, 13)

Davy renders sleep as an uncharted territory of learning and control; from claims of differential consciousness to hypnotic commands to lucid dreaming, the examples characterize sleep as a state we can command—if only we can focus and work hard to do so. The journalist is quick to note that " 'sleep learning' is big business in America, and is growing in Britain, too, with companies offering courses in everything from languages to treatment for nailbiting through loud speakers under the pillow connected to tape recorders" (13). ("Sleep learning" has taken on some different connotations in the twenty-first century, but its potential to be and benefit big business remains the same.)

On the one hand, and even in the face of this potent control, Davy does suggest that we exercise some caution in our labors: "The profound chemistry of the REM episodes, the extreme obscurity of the real nature of sleep and the function of dream, the delicate balance of the three states of consciousness which means 'normal' mental health, all suggest caution and respect for this potent realm" (1967, 13). As with the examples from neurofeedback (chapter 2), there is an early recognition here of the powers that could be unleashed should we begin to manipulate the brain—applying neuroscientific control to those spaces once deemed inaccessible and fundamentally unconscious. On the other hand, intervention is seen as the pay-off at the end of the research. Luce and Segal deliberate on idyllic plans for intervention (plans that have, at least in part, been realized over the past half century): "As we now begin deciphering what its cycles say about our brains and behavior, sleep patterns may become a new Geiger counter in medical diagnosis. Our current progress in biochemistry lends hope that we can develop drugs to supply, not just sleep, but the kind of sleep needed." (1966b, 73).

In the modern age that looked to space travel, already included commuter jet travel as of 1952, and faced the further modernization of the workplace, sleep research promised to offer optimization in the face of disruptions to our naturalized ideas about sleep. "All kinds of modern developments are conspiring to put a stress on sleep habits. Apart from the intangibles of the pace of urban living, modern machinery and computers are bringing increasing pressures for shift work and round-the-clock operation. The sleep habits of large groups of people—especially aircrews and roving businessmen—are constantly disrupted by crossing time-zones in jets" (Davy 1967, 13). Commercial jet travel had only become possible in the early 1950s, and the term *jet lag* was not introduced until the mid-1960s. Yet such perils of modern life have given rise and reason to a set of EEG-based sleep products that promise to alleviate non-normative sleep even under conditions of stress, travel, and the old demon of shift work.

Kokoon and Neuroon: Discourses of Optimization

The Intelligent Way to Sleep   —NEUROON.COM

The origins of experimental sleep monitoring in the 1950s and their uptake in the print news media of the 1960s illustrate a move toward more instrumental monitoring of sleep and its optimization within sleep laboratory situations. In the late twentieth and early twenty-first century, sleep monitoring and optimization moved out of the laboratory, into the home, and onto the bodies of individual users. A number of technologies entered—and sometimes exited—the market over the past fifteen years, including Advanced Brain Monitoring's Sleep Profiler; Zeo's now defunct Sleep Manager; and NeuroVigil's iBrain (discussed in the introduction). Most recently—and successfully—we have two direct-to-consumer EEG wearables: Kokoon, "the world's first sleep sensing headphones" (Kokoon 2017), and Neuroon, "the first intelligent sleep mask" (Neuroon 2015a). Kokoon was developed by Tim Antos and Richard Hall; Neuroon, by Kamil Adamczyk and Janusz Fraczek. Both are available for just under $300; Kokoon expects to finish product development, after years in R&D and a successful Kickstarter campaign, by 2018.

Both devices are polysomnographs, meaning that they measure multiple biophysical signals that can indicate sleep. Each technology includes several of the following measurement techniques in addition to EEG brain wave monitoring: body movement (actigraphy), pulse and oxygen levels (oximetry), eye movements (EOG), muscle tension (EMG), and body temperature. Neuroon relies on three electrodes for its EEG monitoring; Kokoon uses at least two dry electrodes. The reason for combining several technologies to produce a laboratory-like polysomnography effect is best explained by Tim Antos in an interview on the *Sleep Junkies* blog: "We found that using movement tracking or heart rate to monitor sleep patterns just doesn't produce the quality of results needed to actually take control of your sleep, since movement and heart rate during sleep vary widely between individuals." The desire to measure multiple aspects of sleep for discrepancies with ideal sleep points us to a central question: what are the problems that have ostensibly kept us out of control and away from our natural sleep cycles?

As cultural historians and sleep researchers have shown, definitions of natural sleep patterns vary quite widely. Yet, several groups concerned with what has been termed problematic sleep have established national research coalitions. The National Commission on Sleep Disorders Research came into being in 1993; the Association of American Sleep Disorders, which was founded in 1975, changed its name to cover a broader spectrum of sleep issues as the American Academy of Sleep Medicine in 1999. Since 1991 the National Sleep Foundation (NSF) has produced an annual "Sleep in America" poll. In short, the perceived crisis of sleep has established itself on a national stage. David Cloud of the NSF states that the "NSF's most recent Sleep Health index indicated that more than 40 million Americans say lack of sleep or poor sleep had a negative impact on their activities in the week prior to the study. There is a commonly held belief that sleep is optional, so much so that many people have simply forgotten what it feels like to get the right amount" (Mann 2015). Cloud alludes to a sense that Americans are out of touch with natural sleep patterns, that we are so used to "poor" sleep that we have "simply forgotten what it feels like to get the right amount"—whatever that is.

While quality sleep has long been elusive, the problem is currently being framed as endemic among adults in Western contexts. Kokoon executives argue that "we are addressing the common challenge that most adults will face at some point in their lives, achieving quality sleep" ("Startup of the Year" 2016). Neuroon ad copy catalogues similar troubles: "Irregular sleeping times, restlessness, and jet lag are just some of the problems facing professional urban dwellers in the 21st century. The Neuroon gathers data, and with the help of the native app's algorithms, delivers personalized recommendations to the user. Those include optimization of sleeping times . . . and durations in order to use the time available most effectively" (Neuroon 2016b). The *Sleep Junkies* interview with David Cloud partially attributes sleep problems to a factor Alan Derickson identifies in *Dangerously Sleepy*: that sleep is uncool— unmasculine, and unproductive.

> *JM*: Till Roenneberg recently said in the *Huffington Post* that politicians and business leaders who brag about their lack of sleep are like a "modern day tobacco industry." He also said that we have to take away the "uncoolness" surrounding the discussion on sleep. Do you think that technology could help to increase public awareness and turn sleep into something "cool" to talk about?
>
> *DC*: I think Till is on to something; for decades, "burning the midnight oil" has been seen as a badge of honor, further exacerbated by many of the world's most successful people getting by on little to no sleep as they chase success.
>
> But a lack of sleep has significant health consequences, including negatively impacting mood, concentration, memory, productivity and the ability to maintain a healthy weight. That said, technology does have a certain "cool" factor—that's why we see the sleep technology trend as an unprecedented opportunity [to] increase consumer awareness about sleep health. (Mann 2015)

Cloud addresses the larger cultural associations between sleep and laziness, but he also makes a larger point that the opposite is actually true: poor sleep is affecting our health and, by extension, multiple sectors of the American workplace. "We believe that sleep tracking technology will help consumers better understand how their sleep impacts things like mood and performance. They'll discover there is a true sleep duration requirement and they'll see the benefits of making improvements to their sleep habits. These benefits will lead to long-term behavior change" (Mann 2015). As Neuroon's parent company's website indicates, they want to be part of the NSF's solution: "In a long run, Intelclinic wants to partner with organizations, such as the National Sleep Foundation, to deliver scientifically accurate sleep advice based on the readings the mask gathers every night" (Neuroon 2016b). Indeed, Neuroon's marketing campaign is invested in making a phase shift in cultural assumptions and stigmas concerning sleep—particularly around the option of strategic sleep, or naps: "What do Albert Einstein, Nicola Tesla, and Thomas Edison all have in common? They all took naps at work! With the Personal Pause function, you can recharge your mind, and wake up feeling like a genius" (Neuroon 2016a).

Cloud also argues that one does not need to have a sleep problem to benefit from self-monitoring. In the joint National Sleep Association and Consumer Electronics Association survey, Cloud notes, "We did see a contingent of non-users who don't think they need sleep technology, with only 27 percent reporting they believe sleep technology will help them be healthier. Many consumers have the misconception that sleep technology is only for those who have sleep disorders, but everyone can benefit from better sleep. So we see this as an opportunity for the industry to educate consumers" (Mann 2015). Cloud's sentiments illustrate a larger cultural interest in sleeping better and a fear that poor sleep is hurting not only our health but also larger social structures. In short, the joint NSF and CEA report recommends that we get sleep under control.

Initial experiments involving EEG and sleep in the 1950s and 1960s gestured toward control—toward recognizing sleep not as an unconscious state but as a form of consciousness that could be marshaled into a new, natural order. Neuroon and Kokoon extend this logic of technologically enabled control, but they do so through specific technological mediation that cannot be achieved without intervention. One of the taglines on the Neuroon website, "Take Control of Your Sleep," implies conscious, individual jurisdiction, yet the descriptions of the sleep mask demand that users sign over their control to the device. Take, for example, the following descriptions of several Neuroon features, such as Sleep, Bright Light Therapy, Sleep Score, and Neuroon Alarm:

Neuroon Sleep
Sleeping is easy with the Neuroon. Just set your alarm in the mobile app and the Neuroon will take care of all the rest.

Bright Light Therapy
While you sleep, the Neuroon will gradually apply Bright Light Therapy to reschedule your body clock. After a few days of sleeping with the mask you will fall asleep easier and wake up feeling truly rested.

Sleep Score
After you wake up you can check your sleep score in the mobile app to see how you slept.

Neuroon Alarm
The Biorhythm Alarm synchronizes your sleep schedule to your organic biorhythm so you can fall asleep faster and wake up easier. Set the alarm and get your natural groove back! (Neuroon 2016a)

In each case, users are expected to turn over their sleep—its relative ease, schedule, depth, and natural rhythm—to the EEG-based technology. Indeed, users are not even expected to know how they slept; instead, they are directed to a smartphone app that will provide such information. In the popular press, Kokoon is represented as getting us as instrumentally close as possible to the central player in this debate—not the person (via introspection) but the brain—through instrumental measurement: "The EEG allows the Kokoon to directly monitor the part of our body that's actually doing the sleeping" (Boxall 2015).

Central to the claims of control for both Neuroon and Kokoon is, ironically, a conception of nature or one's natural biorhythms. The technology helps to reset something called "your body clock," and it helps you "get your natural groove back." Important to this last claim is the implication that we once had—and have since lost—a natural groove; Neuroon promises to return what we have lost. Kokoon's CEO Tim Antos makes similar claims for his EEG headphones: "EEG sensors measure your brainwaves to accurately track your different phases of sleep—Kokoon headphones can determine the exact moment of sleep, the different phases of sleep and the perfect point in the natural sleep cycle to awaken. This allows the headphones to automatically adjust and tune the audio to ensure that regenerative deep sleep is protected from disturbances and that users wake at the perfect point in the natural sleep cycle" ("Q&A: Kokoon" 2015). In this passage, "the natural sleep cycle" is figured as a fixed feature of the aggregate. Antos's reasoning for using EEG in his headphones (as opposed to actigraphy or EMG alone) rests on the measure's relative uniformity across users. However, Kokoon and Neuroon's "nature" is—as one might imagine—highly mediated and its creation often relies on the "artificial." The Neuroon, for example, "minimizes your sleep inertia by waking you using an 'artificial dawn' (wake up by the light only) and wakes you up only when you are in the light stage of sleep" (Neuroon 2016b). In other words, these

EEG-based sleep technologies rely on what Sarah Franklin, Celia Lury, and Jackie Stacey would call "Nature Seconded," in which "culture becomes the model for nature instead of being 'after' nature" (2004, 194–95).

Indeed, one of the long-term goals of Kokoon is to surpass nature: to employ the technology for mind-hacks, including lucid dreaming, accelerated learning, and even, simply, more efficient sleep. "Our ambition is to give people much greater control of their sleep so that they can get better quality sleep wherever and whenever they need. We would love to be in a place in 10–15 years where people struggle much less with sleep, and are able to get such high quality sleep that they may not need to sleep for as long" ("Q&A: Kokoon" 2015). Such claims echo the hopes of journalists and sleep researchers of the 1950s and 1960s, who noted that "some scientists conjecture that we may shrink our sleep time down to the essential stages, eliminating hours of transitional light stages" (Luce and Segal 1966b 73). Ultimately, the goals of sleep researchers and sleep technologies is not necessarily the preservation of the so-called natural but instead the augmentation and optimization of human sleep for the purposes of efficiency, productivity, and an instrumentally inflected sense of control.

Jet lag (and shift work) applications are, perhaps, the best example of how nature is remade as culture in EEG sleep science. Travel by jet began in 1952; by 1965 the FAA had begun testing the effects of jet lag on circadian rhythms; by 1966 news reporters (Sutton 1966) were using the phrase to describe the phenomenon of moving human bodies swiftly between time zones (Maksel 2008). Although there are plenty of hypotheses about how best to offset jet lag, no "cure" has yet been found. Both Kokoon and Neuroon have made appeals toward jet-lagged users on both ends of the economic spectrum: those who move between time zones as (elite) business travelers and those who sleep in non-normative shifts, including shift workers and airline crews. "While the headphones were originally devised for a better night's sleep at home, Kokoon now believes the market is much bigger than they originally imagined. '1.1 billion frequent travelers and 2.3 billion light sleepers, not to mention snorers, new parents, insomniacs and sufferers of tinnitus are really interested in the product,' Antos adds" (Umapathy 2015). A brief video advertisement focuses on a young woman traveling by train and air who uses the Kokoon headset to sleep peacefully en route.

The Neuroon specifically addresses time-change adjustments with applications such as the Jet Lag Blocker. The description of this feature contains many of the discursive moves we have seen promoting other EEG devices, including deference to the instrument: "Relax and enjoy your flight! The Neuroon will adjust your body clock to the timezone of your destination. The mobile app will also give you helpful tips and advice on how to make the therapy more effective by regulating your exposure to sunlight and recommending the ideal time to sleep. The therapy is applied during the night when you are asleep, so relax and enjoy your dreams, we've got the Jet Lag covered." But the smartphone app images reveal that this "therapy" is played out much like a game: The first screenshot tells the user, "Neuroon helped you beat 38 min of your jet lag" and "Warsaw to New York -7H to beat"; a second screenshot illustrates the user's therapy progress in terms of "biorhythm synchronization" while also making it clear that this therapy occurs before one ever sets foot on a jet plane: "Therapy started 3 days before the flight; Therapy Duration 1 day left" (Neuroon 2016a).

In addition to single-user jetlag, Neuroon claims to be useful for family vacations, "Jet lag can affect adults and children alike. Fortunately, the Neuroon is suitable for all ages! So don't let jet lag ruin your family vacation." For shift work, "Working odd hours? Keeping an irregular schedule? Having trouble waking up and falling asleep? The Neuroon Mask's Biorhythm Adjuster will reset your sleep cycle to your changing schedule leaving you rejuvenated and ready for the day." And for what they call "Social Jet Lag": "Long hours spent in front of the blue light from computer screens (especially at night) have a strong negative influence on our biological clock. The blue light blocks the secretion of the sleep hormone melatonin and interrupts your natural biorhythm. The mask will help adjust to diminished levels of melatonin while you sleep and leave you feeling refreshed" (Neuroon 2016a). In each case, the instrument works by naturalizing the conditions of modern life and synchronizing the user's body rhythms to whatever artificial time construct is necessary for efficiency and productivity.

This intervention of science into the growing social problem of sleep was something Nathaniel Kleitman anticipated. As Alan Derickson notes, it was Kleitman who represented the interests of the American shift worker during the labor debates of the 1940s and 1950s. Among other work on sleep, Kleitman

opposed frequent changes in working hours. . . . The physiologist advised not only individual firms, but the business community in general. The February 1942 issue of *American Business* conveyed his advice to avoid rotation wherever possible and not to change workers' shifts more than once in several months. Kleitman also made constructive proposals for less unsettling nonstandard schedules. He delineated an innovative three-shift plan that improved over the common pattern of eight-hour shifts starting at seven a.m., three p.m., and eleven p.m. Under Kleitman's alternative, workers went to work at noon, eight p.m., and four a.m. This arrangement forced no one to sleep at the worst time of the day, the afternoon. It forced no one to alter normal sleeping time by more than four or five hours. (Derickson 2013, 43–44)

To this problem of shift work, Kokoon and Neuroon offer this solution: adjust the worker, not the schedule. In this respect, "the natural sleep cycle" is only as natural as the business world—and science—deem necessary.

Technology Takes on "a Natural Thing"

The purpose and nature of sleep remain unexplained by science, yet we continue to create technologies that promise better, more natural sleep. Sleep researchers typically acknowledge the disjunction between what we (can) know and what we want to believe about sleep, but "other scientists and laymen may be easily taken with new accounts of sleep findings, especially in the popular press, which point strongly to a brave new theory of the function of sleep. Many such theories look good in isolation . . . Reader beware" (Rechtschaffen 1998, 383).

Kokoon and Neuroon offer us another promise of instrumental intimacy: the opportunity to get (back) in touch with ourselves, this time through wearable technologies that allow us to record, compare, and refigure the states of mind associated with sleep. However, these promising discourses of physiological control belie assumptions about nature-culture, efficiency, and productivity—debates that this chapter calls into question via scientific studies, historical scholarship, and an analysis of print news media. Thanks to the logics of sleep discourses and shifts in definitions of conscious control (see chapter 2), we may no longer sleep simply to rest or to slip away into the unknown darkness of the night; instead, we sleep with and for a purpose: to enhance our natural states of being and reach a state of instrumental intimacy via mechanistic intervention.

4

Neurogeography and the City

EEG's "Collaborative Cartographies"

> My concern with the current turn to new technologies is not with
> the way they might transform a researcher's engagements with the
> world (although that is interesting), rather it is with assumptions
> that new methods or new technologies are necessary to capture
> something that established methods cannot apprehend and
> suggesting that mobility researchers have failed to innovate.
> —PETER MERRIMAN

In "A Future Love Story," author Marcel Van Der Drift introduces us to Steve, a man who is hurling his smartphone into a river. Steve is overwrought: the electronic bio- and psychosensors to which he is "wired" have shown him in graphic detail just how depressed and withdrawn he has become. He had initially resisted the consumer craze of getting "wired up," but after seeking advice from a psychologist, he was persuaded that it would be a good idea to gather some data about himself; as his psychologist explains, "Even adults have difficulty understanding their emotions at some point in their lives. They mostly need feedback from others, but they also gain insight into themselves from a higher perspective, so to say, from objective long-term observation of their behavior. It's no magic, but a useful tool." But one evening, after a fine dinner and (what he assumes to be) a fine flirtation with a waitress, Steve is confronted with her interpretation of their interaction: looking to his phone, "There was the picture of female x, the waitress from Phonsawan restaurant: 'Tired, embarrassed and annoyed when in contact with Steve Smith.' Steve felt a dark, numbing heaviness come over him. 'That's how she saw me. And that's what I am. Annoying.'" This is the final straw, and Steve decides to head for the highest bridge—not to kill himself (although that remains a shadowy possibility) but to destroy his smartphone. En route, he is, ironically, led by the same wired system from which he desires to divorce himself: ignorant of his location or his destination, his shoes vibrate (left or right) to tell him when to turn. "He felt determined as he walked toward the city centre, even though he didn't know where he was going. Being guided by vibrations in his shoes was a little different from the old, long walks to nowhere, but he went with it." This electronic-led walking, overlaid as it is with Steve's older practices of wandering, provides an ideal jumping-off point for this chapter (no pun intended). For, as electronically mediated as Steve's evening walk is, he still recognizes an older practice embedded in the novelty of wired walking: "This was an old habit. At times when he was most down, he'd wander through the city, late at night, looking for certain places. Places he could jump from" (Van der Drift 2009, 29–31).

In chapter 1, I discussed several EEG wearables that publically project one's level of mental arousal to a limited local audience. In this chapter, I discuss two uses for EEG wearables that result in more permanent maps of arousal, reprise our understanding of brain mapping, and aggregate data for political and social agendas. I examine the possibilities and pitfalls of wired walker, bikers, and wanderers armed with EEG devices that purport to record levels of arousal. Often couched in the rubrics of psychogeography, several entrepreneurial and artistic groups have been charting and changing urban spaces through the use of EEG. Wearing bike helmets and neurocams or using smartphones to "see" data installations, creators and users alike are experiencing a conflation of inner and outer space. In a multidirectional and complex mapping, the brain is charted on the landscape and/or the landscape is charted on the brain; either way, the data are central to creating new maps of urban places and neuronal spaces. The maps created by these devices can be limited to the individual or aggregated; in each case, what I have termed EEG "neurogeographies" attempt to reinhabit space and redefine place based on both individual and aggregate arousal.

These maps embody and bring together public displays of arousal, instrumental intimacy, and self-tracking. Although several creators have placed their own projects under the rubrics of psychogeography, we need to recognize the ways that "neuro-" is changing that landscape; using EEG has specific implications for rethinking geographies (of cites and brains). Moreover, these neurogeographies have the potential to aggregate data, compelling us to ask: Who are the data for? What are they expected to achieve? Finally, EEG neuroscapes often depend on discourses akin to those used to describe lie detection: travelers who "report out" via EEG ostensibly produce more accurate data that allow places, spaces, and even brains to "speak for themselves." Throughout the chapter, I question both the idea of novelty—the assumption that "that new methods or new technologies are necessary to capture something that established methods cannot apprehend" (Merriman 2014)—and the notion of "walking with a thesis," promoted by Iain Sinclair.[1]

(Re)mapping a city in novel ways is not a new phenomenon: neurogeographies invoke Guy Debord, who advocated psychogeography through the concept of the *dérive*; at the same time, neurogeographies should be situated within the artistic phenomenon of bio-mapping as it has been advanced by Christian Nold and Stephen Boyd Davis. Wired walking allows us to explore both the older practice of the *dérive* and the contemporary process of purposeful data gathering. Two specific EEG wearables that are being used for neurogeographies are MindRider and Neurocam. MindRider, developed by a quantified self (QS) tracker, Arlene Ducao, is a bicycle helmet fitted with an EEG wearable that records levels of brain electrical activity as its wearer travels about the city. Neurocam, developed by Neurowear, allows users to coordinate an EEG wearable with a smartphone camera; the camera takes photographs of whatever a user is looking at when her or his electrical brain activity reaches an arousal peak. Both were originally intended for use by an individual; however, both have the potential to aggregate data. In the case of MindRider, data sets have already been gathered and published for New York City. While I touch on several other EEG-related (re)mapping projects, Neurocam and MindRider provide an excellent spectrum of neurogeographies.

Defining Neurogeography: Psychogeography, Bio-Mapping, and Plotting the Invisible

As first blush, maps appear to be an objective record of a place: they locate geographic features, demarcate boundaries, and define established routes. However, we know from the work of movement (Mavros 2011; Merriman 2014; Spinney 2015) and feminist geographers (McDowell 1999; L. Nelson and Seager 2004; G. Rose 1993) that mapping is an inherently subjective process, one that can render visible *particular* aspects of place and space. "Despite their apparent message of objectivity, maps select and even distort, because maps, like all images, are made for purposes, and those purposes influence the final form" (Boyd Davis 2009, 40). Indeed, the very things that appear to be the most objective features of maps also reveal their constructed nature: geographic features are selected for their relevance to economic, political, or social value; different maps focus on different features, including geological deposits, tourist sites, or migratory paths. Likewise, boundaries are often national-political borders, not "naturally" occurring divisions, and established routes often favor automobiles rather than, say, pedestrians or cyclists. As such, maps are always already political, subjective, and potentially divisive: they frame the world relative to a particular set of perspectives, create records of the seen and unseen, assume a certain fixity of space and place, and exclude the multitude of interactions between people and their environment that help to shape the world. Indeed, it is perhaps because of, and not despite, these limitations that maps and mapmaking continue to have a powerful effect on how individuals and cultures interact with their environment.

When I use the term *neurogeography*, I have several associations and arguments in mind. Within neuroscientific circles, *neurogeography* has been used to refer to the geography of the brain as an organ being mapped by scientific experimentation (Sharpley and Bitsika 2010). As neuroscientists learn more about the brain, they produce maps—geographies or atlases—that indicate locations and networks associated with certain actions, behaviors, and so on. This aspect suggests ideas of inner and outer space (including, though from a different angle, the concept of "extimacy" from chapter 1). Neurogeography also calls to mind an older term, *psychogeography*, a term that illuminates the mutual imbrication of space, emotion, and behavior for individuals navigating urban environments. Finally, neurogeography reminds us that no matter the interpretation, we are charting physiological data; neurogeographies can therefore be understood as a more specific case of the term and practice of "bio-mapping." As a basic definition, we might say that neurogeographies are impositions of individual EEG data aggregated onto traditional maps, but neurogeographies can also produce their own records that need not be imposed on maps—in this latter form, neurogeographies have the potential to produce more radical interventions, akin to those of the psychogeographers.

Psychogeography and Radical Mapping

One of the most oft cited and ostensibly relevant schools of thought with which to analyze contemporary EEG mapping is psychogeography. As a movement, it emerged in Paris during the 1950s and is often first associated with two avant-garde movements, the Letterists and the Situationists. I leave the larger history and analysis of the movements that created psychogeography to other critics.[2] What concerns me here is both Guy Debord's influential definition and the ways that this term has been—or could be—applied to EEG neurogeographies.

Debord, who was born in Paris in 1931, was an influential member and self-appointed leader of the Letterists and the Situationists. Psychogeography emerged as a term to describe some of the political work the Letterists advocated, particularly in terms of rethinking the city. Debord defines psychogeography as follows: "*Psychogeography* could set for itself the study of the precise laws and specific effects of the geographical environment, consciously organized or not, on the emotions and behavior of individuals" (Debord [1955] 2015, 5). Debord's explanation, which is scientific and experimental, helps to situate a specific type of psychogeography that has resonance in contemporary applications: "the study of the precise laws and specific effects."

> Psychogeography becomes for Debord the point where psychology and geography collide. Gone are the romantic notions of an artistic practice; here we have an experiment to be conducted under scientific conditions and whose results are to be rigorously analysed. The emotional and behavioural impact of urban space upon individual consciousness is to be carefully monitored and recorded, its results used to promote the construction of a new urban environment that both reflects and facilitates the desires of the inhabitants of this future city, the transformation of which is to be conducted by those people skilled in psychogeographical techniques. (Coverly 2006, 89)

These descriptions of psychogeography in theory and practice resonate with some of the underlying themes of the artistic and scientific bio-mapping projects under way. Neurocam's Neuro Tagging Map and MindRider, in particular, take on the task of merging electrical brain activity and geographic space, so much so that the city maps in *MindRider Maps Manhattan* are reminiscent of brain-imaging visualizations.

Importantly, the spirit of the larger Situationist movement had a revolutionary, radical flair, as either "oppositional" to the authority and strictures of city planning or as military mapping for invasion; walking held a special place in this constellation as a radical means of changing and challenging city spaces: "Walking is seen as contrary to the spirit of the modern city with its promotion of swift circulation and the street-level gaze that walking requires allows one to challenge the official representation of the city by cutting across established routes and exploring those marginal and forgotten areas often overlooked by the city's inhabitants" (Coverly 2006, 12). This type of wandering reveals new things about urban spaces—the kinds of political, social, economic, and political factors that divide cities, rendering certain regions more or less desirable than others. Technologies such as MindRider are described in similar terms. Arlene Ducao argues that "the process of examining the current MindRider Manhattan Map inspires us, for even with its modest scope, it unveils patterns that reveal the city in an entirely new way" (2014, 57). For Debord, "psychogeographic articulations of a modern city" reveal certain "pivotal points": "One measures the distances that actually separate two regions of a city, distances that may have little relation with the physical distance between them. With the aid of old maps, aerial photographs and experimental *dérives*, one can draw up hitherto lacking maps of influences, maps whose inevitable imprecision at this early stage is no worse than that of the earliest navigational charts. The only difference is that it is no longer a matter of precisely delineating stable continents, but of changing architecture and urbanism" ([1956] 1981, 66). Debord leads us to an important concept: that as psychogeographers we are not looking at or for stable contents but instead attempting to trace out a record of the instability of (urban) spaces—not only the ways they shift over time but also the ways that they are inhabited, understood, used, and felt or experienced.

In their radical reconsideration, *dérives* share an opportunity for reterritorialization that has been associated with military remapping. "It has been claimed that, far from being the aimless empty-headed drifting of the casual stroller, Debord's principle is nearer to a military strategy and has its roots not in earlier avant-garde experimentation but in military tactics where drifting is defined as 'a calculated action determined by the absence of a proper locus.' In this light, the *dérive* becomes a strategic device for reconnoitering the city, 'a reconnaissance for the day when the city would be seized for real'" (Coverly 2006, 96–97). The references to military tactics and seizure should remind us that mapping is not an innocent activity, here or in the case of EEG neurogeographies.

And yet, the Situationists' and Debord's impacts on radical, political change are less than clear. In addition, Michel de Certeau argued that voyeuristic objective "from above" looking at the city erases the sense that individuals and the local matter and are present there, yet a contradiction is at work: "A map can never accurately capture the lives of those individuals whose journeys it sets out to trace for, in the process, individuality is inevitably flattened out and reduced to points on a chart"; or put another way, "the objective and programmatic approach of the sociologists and geographers threatens to obscure that which they seek to preserve, rendering them less able to assess urban life accurately than the historians and novelists that have preceded them in this account"

(Coverly 2006, 106–7). This idea is echoed in Justin Spinney's essay concerning the more technologized practice of bio-mapping: "There is the very real danger that using bio-sensing in this way may do exactly the same for affect as Vergunst has cautioned: 'turning too readily to high technology has the danger that we actually distance ourselves from the experience of movement, in the very act of trying to get closer to it'" (2015, 241). In this respect, some of the drawbacks of psychogeography continue in modern incarnations of "walking with a thesis."

In many facets, psychogeography informs contemporary practices of neurogeographies: from Debord's original definition to the extant maps' potential political and social ends to the potential alienation of tracking rather than experiencing movement. Yet as we move from the Situationists' experiential records to biological data and eventually to the brain's electrical activity, there are some new considerations to which we must attend.

Bio-Mapping and Plotting the Invisible

In addition to psychogeography, bio-mapping is an important antecedent to neurogeographies. The push to create mapped collectives of various bio-data has emerged across several disciplines, with geography and the arts at its center. Indeed, over the past two decades, bio-mapping has become a worldwide phenomenon (Nold 2009), and while it does have precedents, imitators, and interrogators, the bio-mapping project to which many contemporary projects refer began in 2004 with the work of Christian Nold. According to Nold's introduction to *Emotional Cartography: Technologies of the Self*, "Bio Mapping emerged as a critical reaction towards the currently dominant concept of pervasive technology, which aims for computer 'intelligence' to be integrated everywhere, including our everyday lives and even bodies. The Bio Mapping project investigates the implications of creating technologies that can record, visualise and share with each other our intimate body-states" (2009, 3). Nold's projects require participants to wear both a sensor that measures galvanic skin response (GSR) and a GPS device that records where the GSR arousal waxes and wanes. The participants then take walks around a particular city or locale while their GPS and GSR data are recorded; the devices do all of the work, and participants are not required to interface with the technology while walking. In this respect, the GPS-GSR combination functions like a polygraph: gathering data that the subject is unable to consciously control. Nold then collects the data; debriefs participants, asking them to provide introspective reports about their journeys; and aggregates all of these data into maps that he freely distributes to the local community under investigation (figure 4.1).

Nold's projects are intended to challenge the world in which Marcel Van Der Drift's protagonist Steve found himself immersed. Yet by wiring walkers with bio-mapping technologies, Nold's project also enacts the same techno-centric world he wishes to critique. This irony is not lost on Nold, and his aim is to "imagine the social, economic, cultural and political implications of creating a public Emotional Cartography" (2009, 7). The essays in his collection bring this thought project into relief, leaving readers to make their own judgments about living in a world of public displays of arousal.

Nold's projects, and bio-mapping more generally, fit into a larger movement of tracking, cataloging, and changing the ways that people experience their locale. For example, bio-mapping jibes with the quantified self movement (see the introduction). The difference is that after individuals track themselves, their data are aggregated by bio-mappers to create representations of place that illuminate gradated differences in whatever variable is being tracked. Here we might also include related projects such as "Pulse of the Nation," a collaborative project that charts US "mood" based on Twitter posts, and "We Feel Fine," a web-based project that "has been harvesting human feelings from a large number of weblogs" since 2005 (Harris and Kamvar 2006).

One of the latest trends in bio-mapping is to apply EEG wearables to the task of "arousal" tracking: some examples include an architectural walk study conducted in Edinburgh (Aspinall et al. 2013) and "Conductar," an installation at Moogfest in 2014. "Conductar: Moogfest is an immersive augmented-reality experience, layered on top of the entire city of Asheville. Moogfest goers are invited to wander the city and conduct a generative audio-visual world through movement and their neurological response to the environment. Inspired by the psychogeographical concept of 'Dérive,' the installation runs on a mobile app connected to a brainwave sensor. As festival goers drift from one venue to the next, they will collectively compose new electronic music with the electrical activity of their brains (EEG data)" (Conductar.com 2015). In this example, connections to psychogeography and the *dérive* are explicit; the difference is in the method of recording. Whether the technological interface adds to or changes Debord's concept has yet to be determined; however, both Aspinall et al.'s study and Conductar: Moogfest intend to modify our understanding of urban spaces by recording and displaying our reactions to wandering the city streets.

Figure 4.1 (above and opposite). Greenwich Emotion Map. From Christian Nold, *Emotional Cartographies* (2009). Used with permission.

One final neurogeography that should be included in this growing pantheon is "Things We Have Lost," the brainchild of John Craig Freedman and Scott Kildall (Freedman 2015). The project was funded by the Los Angeles County Museum of Art's Art+Technology Grant in 2015 and has thus far included an installation in Los Angeles and one in Liverpool, England. The project began in two stages: first, participants were brought into the lab to have the electrical activity in their brain recorded as they concentrated on something they had lost.

GREENWICH EMOTION MAP by Christian Nold

How was the Greenwich Emotion Map created?

What is the relationship between emotions and physical space?

How can you use this map?

www.emotionmap.net

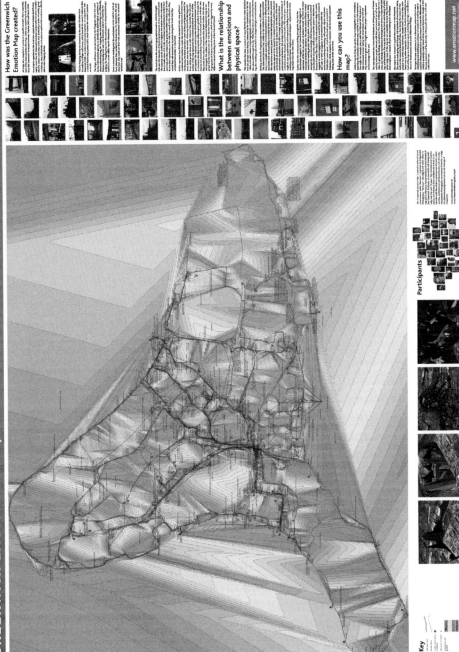

Participants

Key

Scale 1:3530

The second portion of the project involved interviewing participants on the street and posing the question: what have you lost? Later, these objects were programmed into GPS so that materialized versions of the lost objects could be "set" on the map in these interview locations. Now, people who have downloaded the app to their smartphone can wander the city tuning in to these objects and seeing them with their smartphones. Bio-mapping projects resonate with the potential for neurogeographies to alter our experiences of space by inserting digital objects, which cannot be seen with the naked eye, into the landscape.

But several issues should inform the critical analysis of bio-mapping data. First, what are we choosing to record? Humans have long sought to record or make visible what was once unseen, from germs to radio waves to nervous electricity. Advances in technology allow us to "see" and document phenomena that were either previously unknown or at least out of our spectrum of perception—including the electrical activity produced by our brains. Whenever we record or map such a phenomenon, we make choices about what is worth representing, what representations are technically possible, and which of these will be presented to users. While we could characterize these visualizations as discoveries, it would be more apt to term them constructions, given that as one records, one also categorizes, compares, and fixes the latest visualization in a larger constellation of knowledge making (Foucault 1995, 187). Once rendered "knowable," these new objects of study often reflect not something inherent in their nature but many of the cultural assumptions into which they are born (see, for example, Latour 1987 and Schiebinger 2004). Whether it uses GSR, ECG, EEG, or other measurement mechanisms, bio-mapping brings the construction and "fixing" of the unseen into relief.

A second, and related, issue is the acceptable and desirable mechanism(s) for collecting physiological data. Here, I use the term *physiological* because it better captures the changeable nature of the phenomenon being recorded. Like Debord's *dérives*, physiological measurement does not capture a stable architecture of the body. Because physiological data often involve the autonomic nervous system, it is important to consider the values placed on collecting these (sub- or unconscious) data. Since the early twentieth century, and the rise of brass instrument psychology, *how* we measure has come to matter; specifically, automatic measurements have gained credence over introspective measures in the laboratory. The rise of lie detection depended on and was informed by this same logic: that the autonomic physiology of the body will reveal more than a subject can or will consciously convey (Littlefield 2011). Similarly, contemporary mobile EEG bio-mapping incorporates conscious introspection, but this is in addition to the collection of autonomic data and patterns of the brain's electrical activity. As in lie detection, a subject could concentrate on a mathematical problem, close her eyes, or focus on stressful thoughts to modify the physiological rhythms of her body; likewise, mobile EEG can capture data from subjects that are less mediated and ostensibly therefore more objective. However, we know from scholarly work in the fields of science and technology studies (STS) and literature and science that objectivity is constructed (Daston and Galiston 2004) and that it is often defined in relation to, and so cannot be fully distinct from, subjectivity (Squier 2004). Thus, these "objective" bio-data are constructed, as are the maps onto which they are charted.

We must also be aware that the study of any specific task or physiological factor requires comparison to a control or an atlas of recognized possibilities. As we saw in chapter 2, EEG studies often rely on previously established "brain maps," or targets. As with a target heart rate, which Stephen Boyd Davis uses in his bio-mapping project " 'Ere Be Dragons," the expected ranges are established ahead of time and typically in relation to one of two standards: either in comparison to an individual's resting heart rate or in comparison with an established chart of aggregate heart rates based on age and fitness level. Likewise, in the case of EEG, the bio-mapper must decide on a baseline and a target. As we have already seen, many of the mobile EEG makers do not release their algorithms for determining such targets. We can speculate that the levels of "arousal" being measured are based in relation to both a control sensor (often worn on the earlobe) and standardized EEG target wavelengths, which are then recoded with names such as "alpha," "beta," "theta," and so on, and associated with particular states of mind that can be rendered more or less desirable depending on the situation. At issue are not only the constructed nature of the data's representation (as we saw in chapters 1 and 2, the labeling of any given electrical activity as of a particular wave or as "expert" creates predetermined expectations and benchmarks) but also artifacts and noise created by an inefficient system of collection.

Representations of data determine why we would be measuring something like "arousal" in the first place and placing any value on it as a tool for self- or social understanding.[3] Nold and others have illustrated that we are capable of measuring "arousal" via multiple, autonomic physiological processes (such as GSR or ECG); why then should we turn to the brain? What is of interest and value to the more recent mobile EEG researchers, such as those working on Neurocam and MindRider? Is there something new here—beyond even connections to psychogeography and bio-mapping? The answers can be partially found by examining two EEG technologies for creating neurogeographies.

MindRider: Recording the Mental Signature of a City

When Arlene Ducao began combining bike helmets and portable EEG headsets, her goal was to create a visualization of brain activity for riders traveling through congested city streets. The first prototype of MindRider, a bicycle helmet that collects EEG data from a user via a portable sensor and combines it with GPS and other data to create alternate maps of cityscapes, was created at the MIT Media Lab. Later models, designed by Ducao and Ilias Koen, were produced by DuKorp at the Brooklyn-based DuKode Studio. Together with a larger team, including Josue Diaz, Ducao completed her first collective map, *MindRider Maps Manhattan*, in 2014. Ducao has categorized MindRider's maps as "psychogeographies" ("MindRider and the Maker's Brain" 2014) and in some respects, her team's work fits Debord's definition. However, there is more to MindRider than psychogeography. MindRider is redefining relationships between the brain and the city, particularly through discourses of communal mapping. Using EEG, MindRider generates cartographies that are both communal and collaborative, whose discourses embody both the aggregate and the personal. *MindRider Maps Manhattan*, the project's first publication, overlays neurodata and public works data, suggesting new potentials for politicizing how the brain relates to the city.

MindRider could be classified among bio-mapping projects aimed at aggregating and mapping emotion, affect, and arousal. These projects have been characterized as creating a "communal emotional surface" (Nold 2009, 7); fashioning "collaborative cartographies" (Mavros 2011, 32); or producing "affective geographies" (Giaccardi and Fogli 2008). MindRider picks up on this trend in order to provide maps of a city based on states of mental arousal. Notably, MindRider's advocates describe the technique using discourses in which aggregation is assumed not only between participants but also between brain and city. In online advertising and in various presentations, Ducao and others refer to MindRider data as "mental snapshot of the city" (DuKorp 2015) or a "mental signature of the city" (DuKode Studio 2015). By invoking the photograph and the signature, they evoke traditions of recording potentially individuating human forensic features. But in this strange inversion, the city—not its neurogeographer participants—is granted individuality thanks to brain-based data. From over a century of forensic work, we know that signatures can be forged and photographs can be altered—that individuality according to certain metrics remains in question. In this respect, the mental signature of a city or its mental snapshot captures fluidity rather than stability, yet what remains intriguing is the desire to attribute "mentality" to a place.

Aggregating physiological data and city maps is a common point of inquiry for other bio-mappers. In Stephen Boyd Davis's experimental analysis of a project using GPS and ECG, he argues that "players see themselves from outside, as though they were distant elements of the cityscape; they also see a process from their own interior space combined with landscape data" (Boyd Davis et al. 2007, 23). Christian Nold made several different kinds of maps using GSR data, each aggregating, blending, and applying the data to the city in different ways; each reveals the power of choice and intention in the creation of representations. Take, for example, Nold's description of one of his many "emotion maps": "in the Greenwich Emotion Map, this meant using a GIS (Geographical Information Systems) software to create a communal arousal surface which blended together 80 people's arousal data and annotations. The resulting communal 'emotion surface' is a conceptual challenge and question. Can we really blend together our emotions and experiences to construct a totally shared vision of place?" (2009, 7). Nold's main question implicates the blending of data as the primary issue, but his experience also raises the specter of mapping humans onto place.

If the larger, aggregate data descriptions are telling, so too are the discourses used to describe smaller "moments" that are also spaces and places within the larger map or city. MindRider is expected to document "your experience and engagement in real-time," and it does so through an algorithm that records your relative arousal, translating that arousal state into a color (see chapter 1)—framed in the color language of the familiar traffic lights that litter the city, red implies "pure attention" and concentration and green indicates "pure relaxation" and calm. Indeed, Ducao mentions that MindRider functions within the "vocabulary of traffic lights" ("MindRider and the Maker's Brain" 2014), yet the red, yellow, and green colorways are not intended to instruct riders during their commute.

Interestingly, another further translation also happens at this point: MindRider proponents refer to the red zones as "hotspots" and the green zones as "sweetspots," creating an implicit evaluative scale for the riders and mimicking some of the colorization often used in brain imaging. While this discursive move may seem innocuous, it can be used both for training purposes (to train the rider's state of arousal over time) and for city planning purposes (to change the hard and soft infrastructures of the city). Both potentials again bring brain data and the city into a strange—but potentially productive—confluence.

First, MindRider's discourses imply a potential for individual change. As we saw in chapter 2, EEG feedback is being used for the training of elite athletes; likewise, those who want to simply sleep better can use EEG wearables such as Kokoon and Neuroon (chapter 3). In the case of MindRider, we are told that the data are useful as a means "to build a better ride for the future" (M. Craig 2014). This idea of using MindRider data to build something better is repeated on the project website: "This insight provides the opportunity to challenge, enjoy and maximize your experiences." Specifically, we are told that "one Mind-Rider cyclist turns his commute into a kind of mind-training game. After reviewing his MindRider maps and learning which routes have the most Sweetspots and Hotspots, he often challenges himself to relax his mind and turn his Hotspot-heavy routes into Sweetspots all the way from home to work" (DuKorp 2015). However, as we also saw in chapter 2, EEG training tends to involve direct neurofeedback and must be practiced over time if one hopes to match a particular wave structure. The MindRider helmet does not provide this kind of support; thus, the user-based claims concerning training have more to do with self-perceived stress levels than brain-activation data or the technological support provided by MindRider.

The project also suggests potential for political (aggregate) change. Mind-Rider data are being collected from individual riders within specific cities, aggregated, and combined with other infrastructural data sets. The idea is that data taken directly from our brains during a commute could provide insight into how and where the infrastructure of the city should be modified. In *MindRider Maps Manhattan*, Ducao makes concrete suggestions for infrastructural monitoring and potential policy change. For example, when correlating "Vehicle Collisions and Bike Routes," Ducao argues that the "findings suggest that the Downtown zone be further surveyed for improved planning interventions, especially where collision occurrences increase near bridges and tunnels" (2014, 53). This reasoning comes from a particular amalgamation of data sets that bring individual user data into contact with public, municipal data.

MindRider Maps Manhattan plots MindRider data on top of or alongside NYC Open Data repository information, including crash reports, traffic violations, and construction, as well as zoning maps, and environmental protection reports. To create the maps, MindRider data and NYC data are presented as "mash-ups" (Ducao 2014, 8) in which one data set is overlaid on another data set. The maps are not labeled with street names or any other indicators that typically adorn maps; moreover, the island of Manhattan is literally rendered as a floating compilation of red, green, and grey dots (figure 4.2). Particular areas of interest are enlarged and accompanied by textual explanations: "pothole complaints are so abundant and so evenly distributed throughout Manhattan that they don't seem to filter any significant features in the MindRider data" (25), or "noise complaints overlap most densely with MindRider data in the northernmost and southernmost parts of the island—where Manhattan's streets and houses are least gridded" (27). While the discussion of statistical significance is limited to the end of the book, the colored, dis-located, unlabeled map enlargements serve as impressionistic portraits of a city replete with untapped data flows.

The impressionistic nature of *MindRider Maps Manhattan* is also reminiscent of brain-imaging data, particularly as they are represented in lay publications. The MindRider data are coded from red to green, while the municipal data are represented in various shades of grey. At times, and particularly in the map magnifications and the section on zoning, "MindRider+Planning," the MindRider data light up the grey, disembodied, and dis-located city much like the red, orange, blue, and green activation markers color otherwise grey brain scans. In this respect, the city is, itself, transformed into a brain image, one that has data flowing through it, one that is electrified by sweetspots and hotspots. During rare moments when city streets and bike routes are marked as grey squiggling lines, they most closely resemble a jumble of nerves that have particular concentrations of red and green activity. In each case, neurogeography has rendered the city and the brain as mutually informative and, perhaps, indistinguishable.

The process of mapping the arousal levels of riders and correlating these data with other data sets reminds us of mapping the unseen and the aesthetics of transparency. The mapping process might also recall the ideals of studying expert brains and using those data to remap the brains of novices. MindRider, however, adds another element to the ideas of charting, transparency, and expertise embodied in EEG wearables: the base maps of the city are assumed to be the objective, physical structure that is already in place. Onto that framework, Ducao and colleagues place subjective-objective data in order to make

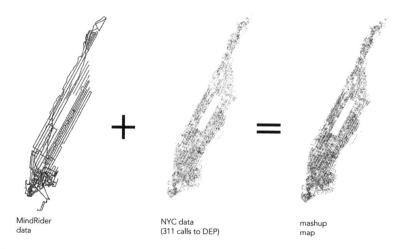

MindRider   NYC data   mashup
data   (311 calls to DEP)   map

Figure 4.2. MindRider Maps Manhattan data overlays. ©2015 by DuKorp. Used with permission.

several political statements about traffic safety and Manhattan.[4] Of course, as with most maps, these show us only the basic outline of city roads and routes, not the condition of the routes, the weather, or the history of the routes. So they are always already also displaying an absence of information that can only be obtained through supplements. The question then becomes which supplements are the best, the most adequate? And why might brain data be more persuasive than traffic reports or other types of data sets—bio-physiological or municipal? One tentative answer may simply be that in their current form, these brain data sets are most valuable to the individuals who use them as a commuter aid. In discussing her debriefs with MindRider participants, Ducao notes that "after many of their journeys, riders used the resulting maps—their mental snapshots—to explain their experience to me. I came to realize that each rider's map was a kind of quantified memory aid" (2014, 57). And it is to this particular use of EEG wearables as a quantified memory aid to which we now turn.

Neuro Tagging Map and Neurocam: Recording a Documented Life

Neurocam was first introduced in 2013 by Neurowear, the same company responsible for the Necomimi headset (chapter 1). Neurocam has no "ears" to project one's states of arousal outward, but the device does include a headset with a smartphone mount so that it can take photos via the phone's camera. "The hardware is a combination of Neurosky's Mind Wave Mobile and a new customized brainwave sensor with the newest BMD chip. It transforms the smartphone into a brainwave analysis device + camera with the smartphone screen displaying scenes of interest as they are being recorded" (Neurowear 2015b). Neurocam can be used autonomously or in conjunction with another Neurowear product, the Neuro Tagging Map.

Neurocam and Neuro Tagging Map simultaneously serve as territory-marking devices and as personal memory aids. The communal aspect of many psycho- or neuroscapes is not entirely lost in these applications, but the goals and potential for sharing are distinct. Moreover, and unlike MindRider, these products are prototypes and not available to a wider market. Therefore, my sources are limited to advertising campaigns, initial convention presentations, and ideal and sometimes fantastic applications. In this section, I introduce Neuro-cam and analyze its public presentation, paying particular attention to the tension between self-knowledge and the potential for revealing states of arousal to a larger, more public audience. Next, I examine Neuro Tagging Map and its potential for marking territory. Finally, I compare Neurocam and Neuro Tagging Map to other—as yet—imaginative devices that could serve as memory aids that account for time and space, as opposed to simply place.

Neurocam: Self and the City

A video advertisement for Neurocam begins with a brain broadcasting to an icon of a video camera; the catchphrase is "remember your moments." From the outset, the brain is being represented as an inadequate memory aid. In the next scene, a young woman wearing the Neurocam stands outside on a beautiful day with her eyes closed. She attaches her smartphone and opens her eyes. The phone's readout goes from 15 to 45 (a sudden change that occurs as alpha waves give way to beta with the mere opening of the eyes). The advertisement then follows the young woman through various activities and interactions: seeing a pastry (which registers a 73), following a small dog (which registers a 69), and looking at a toy rocket ship (which registers a 71). The numbers after each example are displayed on her phone; only by reading the ad copy do we learn that this is because the phone and headset are programmed to work together: when her brain's electrical activity reaches certain states of arousal (which Neuro-cam translates as anything in the range of 60–100), the phone's camera snaps a picture so that the user can remember the event, thing, place, or person. The video ends with the young woman sitting down to have coffee with a young man. They amuse themselves by looking through her (arousal) photos for the day. They share a laugh over most of the pictures and display embarrassment when they reach the final photo, a Neurocam picture of the young man himself, which registers a 99.

Unlike MindRider, Neurocam is not—at least at its initial marketing—intended to be used for communal cartography. It does have shared components, but the primary purpose is individual: memory keeping and geographic marking. We are not privy to the algorithm that converts the brain's electrical activity into a ranking of 0–100 points; what we do know is that the device must register 60 or more points to take an automatic photograph of whatever is in front of the wearer. The photos are then stored for later retrieval. Like Mind-Rider, the concept is simple—in this case, an amalgam of camera and EEG—but the idea of recording the otherwise "unseen" removes the agency of the user from the equation. One does not need to have a camera at the ready to record things, places, or people of interest; instead, the camera can "sense" what is of interest, even before the user does. In this respect, Neurocam seeks to use one's nervous system to reveal details about one's desires, arousal points, and so forth. The idea is that the camera's photo roll will reveal something to the user (and to whomever she shares the images with). The difference between MindRider and Neurocam may simply be with whom the data are automatically shared.

Here we have the recurring problem of objectivity and subjectivity wrapped into a seemingly innocuous camera-EEG set-up: subjectively speaking, if we like something, we might take a photograph (or simply commit it to memory); conversely, there may be things, places, and people we do not register (even if they are important or arousing to us). With Neurocam, we are able to capture, "objectively" speaking, all of the things that raise our brain waves to a certain arousal point: anything above 60. Yet, as Stephen Boyd Davis argues regarding another bio-data GPS application, there are always already multiple levels of subjectivity at work even in the most "objective" data gathering. His arguments remind us that subjectivity is not simply about "likes" and "dislikes" or our own individual preferences; rather, subjectivity is always already informed by larger cultural mores and expectations *and* by the technologies we use to record or capture what we deem important. And objectivity is indivisible from subjectivity (as feminist philosophers and theorists of science have argued for decades).

To represent is to select, interpret, translate, transform: this is the first level of subjectivity. Next, selections and distortions which belong to whole cultures and to groups and subgroups are evident—these are the cultural subjective and subcultural subjective. The latter may take on a conscious socio-political aspect, as in feminist geography, or may simply reflect in some way that the world is different depending on who you are and where you start from. Interactivity introduces the possibility of customisation to the individual and to the time: the individual subjective both spatial and temporal. With the advent of portable interactive technologies sensitive to multimodal inputs, what matters to me now at this moment in my current location and circumstances can become central. This introduces a form of subjectivity we could call the egocentric subjective, in which my location and other aspects of myself impact decisively on the representation. (Boyd Davis 2009, 49)

Boyd Davis would argue that these new mobile technologies *are* having an effect on the kinds of data that can be collected *and* on the kinds of conclusions or positionalities available. But this "egocentric subjectivity" is not novel—indeed, it returns us to the *dérive*. Boyd Davis's "'Ere Be Dragons" "revisits such mapmaking. The parts of the territory which are mapped, and the visual form that the mapping takes, are subjective in the double sense that mapping only occurs for the immediate locations through which the interactor passes—it is a record of a personal journey, like the periplus, historically a sequential list of features encountered as one navigated a coastline" (Boyd Davis et al. 2007, 22). More than MindRider, Neurocam is well suited to capture egocentric subjectivity, but it's the liminal point at which the self and the map collide that makes this technology truly intriguing.

Neurocam Meets Neuro Tagging Map

Neurocam is interesting in and of itself, but when paired with Neurowear's Neuro Tagging Map the combination reveals a technological system akin to MindRider: the addition of communal-style GPS to mark maps for others to see as the smartphone app tags the place with a numeric value. The makers of Neurowear have likened their Neuro Tagging Map to a primitive instinct, ironically using digital technology to bring users back to instincts of place and marking they have lost. "Many animals leave biological information such as horn/nail-marks, body odor, body waste etc. in their living environment to mark their territory or share information such as their health status. It is an important form of animal communication that human beings lost over time. We asked ourselves if there wasn't a way to leverage the power of technology to bring this form of instinctive communication back to humans. The smartphone app 'neuro tagging map' offers a new way of communication based on location information and biological information" (Neurowear 2016). The idea of categorizing brain data as biological information akin to even "horn/nail-marks, body odor, body waste" renders the brain part of the body. The implication is that Neuro Tagging is instinctual—an analogy that Neurowear plays with in its other products, including the cat-eared Necomimi cosplay headset. The reasoning behind using this idea of instinct could very simply be that Neuro Tagging is automatically recorded, or, more to the point, that our brains are always already marking places, people, and things (with or without our conscious intention) and that this mental "tagging" is and should be something that is communally shared because it is helpful for the community. What is strange, however, is calling this a "new form of communication"—put another way, we might say that a new technology has simply made this already existent form of instinctual (but until now invisible) form of communication visible.

Because Neurocam and Neuro Tagging Map are in development, we do not have many descriptions or images to work from. But the artist's rendering of the Neuro Tagging Map includes images and data added to an extant map. As Stephen Boyd Davis and others have detailed, the act of combining images and maps has many precedents. Several bio-mapping projects have created or overlaid maps with images. Christian Nold's bio-mapping projects have produced various "emotional cartographies" for several cities. While some of his maps use color and shading as a means to represent the emotional (in his case GPS) data, others rely on adding images and pictures to re-draw maps of cities, such as Stockport (figure 4.3). These maps not only include GPS data and "emotional arousal clusters" (rendered as red cylinders), but also images drawn by the participants and linked to specific places (Nold 2009). As with a Neuro Tagging Map, these images embed and make obvious the subjective data in any mapping project as well as revealing the ways that "objective" maps omit relevant data. These information additions might be suitably compared to MindRider's interpretive "sweetspots" and "hotspots."

This practice of annotation appears to fit a larger trend in geography, as Panagiotis Mavros notes: "In the recent, yet widespread, practice to annotate information on locations and leave it available for others or, in reverse, to browse through others' annotations, Space is understood as an infinite archives, a repository that should somehow be updated and retrieved according to the desire of its users. This perception of space encourages a new kind of sociability and enables a new collective understanding of places, proposing a new way to learn 'about each other's relationship with a place'" (2011, 31). Mavros's sense of the collective and of repositories matches well with Neurowear's descriptions of Neuro Tagging Map: these maps are intended to share a different kind of information in addition to their typical use as wayfinding devices: "neurowear believes that neuro tagging brings a new value to the places being tagged" (Neurowear 2016). Like "Things We Have Lost," the tagged map serves as a lasting reminder of the ghostly traveler who once walked these roads.

Neurogeographic Memories

Neurocam and Neuro Tagging Map posit that "remembering one's moments" is not experiential but tactical: memory is reduced to storage, the potential for over-mapping data, and the opportunity for living not in the moment but vicariously through our own records. Looking back at these data, as the young woman in the Neurocam advertisement does, allows one to revisit one's *dérive* and potentially see things that one missed in the moment. For the woman, it is a surprise—and an awkward moment—when she flips to the image of the man sitting next to her. Creating neurogeographic memories, then, is as much about algorithmic data unconsciously extracted from users as it is about conscious, post-experiential interactions between a user and her data.

Because the Neurocam is a prototype, we can only speculate about its future use; through MindRider, we can imagine some of the consequences of this distancing and experiencing through wearable neurotechnology. In a *New York Today* article concerning MindRider, the authors note that "after dismounting, cyclists can also check an app to see the data from their trip plotted on a map." One rider reflects, "Once a woman opened up her cab door, and I collided with the door. I remember thinking, I can't wait to see this" (Correal and Newman 2014). In this brief quip, we are privy to the user's awareness that his device is creating loggable, revisitable memories for him to experience (again) later.

Indeed, in each of the bio-mapping projects discussed in this chapter, the individual's post-experience interaction with data matters. For Boyd Davis, Nold, and the creators of Neuro Tagging Map, reflecting on the data—perhaps seeing them for the first time—and reliving the experiences of the *dérive* are additive, experiences in and of themselves. "It is remarkable to look at the diversity and uniqueness of an individual's experience of the city. Some people's responses

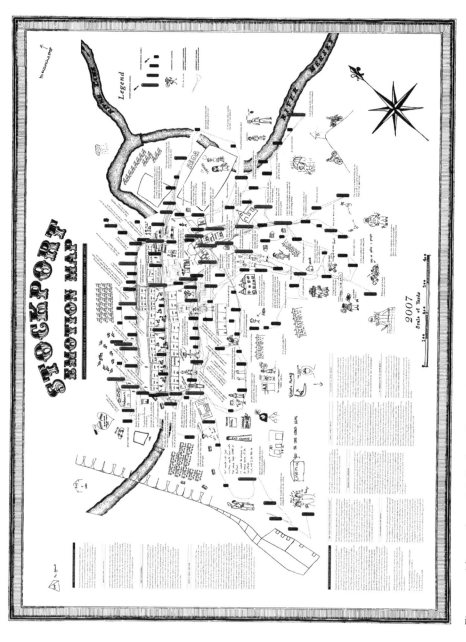

Figure 4.3 (above and opposite). Stockport Emotion Map. From Christian Nold, *Emotional Cartographies* (2009). Used with permission.

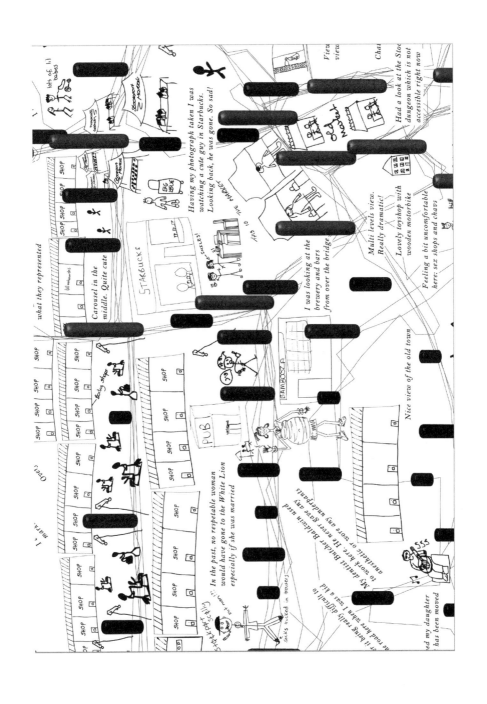

lots of (i) babes

what they represented

Carousel in the middle. Quite cute

Overy...

I miatra...

SHOP SHOP SHOP SHOP SHOP SHOP SHOP SHOP SHOP SHOP SHOP SHOP SHOP SHOP SHOP OP

STARBUCKS

PUB

Baby shops

In the past, no respectable woman would have gone to the White Lion especially if she was married

Socks tucked in houses

Having my photograph taken I was watching a cute guy in Starbucks. Looking back, he was gone. So sad!

MARKET

art in city

I was looking at the brewery and bars from over the bridge

BAMBOO

My dentist Butcher Baldwin used to work here. He never gave any anaesthetic or more any underpants

...ed my daughter has been moved

...er it living really difficult to ...he road here when I was a kid

old market

Had a look at the Sto... dungeon which is not accessible right now

View... view...

Cha...

Multi levels view. Really dramatic!

Lovely toyshop with wooden motorbike

Feeling a bit uncomfortable here, sex shops and chavs

Nice view of the old town

are shaped by their memories while inspiring views, old Victorian houses, or green spaces influenced others. There are still others who responded by absorbing the present" (Nold 2009, 37). Here, again, we sense the presence of Debord and the *dérive*, with the exception of the radical agenda, which has fallen away in favor of personal experience: a Pilgrim's Progress "in which geography becomes a surrogate for events in the traveler's personal development" (Boyd Davis et al. 2007, 22).

It is as if science fiction writer Philip K. Dick himself wrote this chapter in our wired history; indeed, Dick has long been interested in issues of (altered) memory, and in "We Can Remember It for You Wholesale," he directly addresses the problems of memory through Douglas Quail, a man who cannot remember if he has really been to Mars. In a meeting with an agent from the Rekall Corporation, we learn about the fragility of human memory: "Had you really been to Mars as an Interplan agent, you would by now have forgotten a great deal; our analysis of true-mem systems—authentic recollections of major events in a person's life—shows that a variety of details are very quickly lost to the person. Forever. Part of the package we offer you is such deep implantation of recall that nothing is forgotten" (Dick 1997, 308). While there is much more to the story and the philosophical issues that it raises, Dick's story helps to explore the consequences of corporate interventions in memory making. Likewise, we might look to *Final Cut* (Naïm 2004), a science fiction film starring Robin Williams, in which individuals can be implanted at birth with a chip that records their life as a nonstop filmic event. Once someone dies, a "Cutter" takes all of the data acquired over the life span and edits it into a montage that can be played back at a memorial service. In the filmic world, having the chip is considered a desirable statement about one's socioeconomic position; only people of the lower classes and the Cutters themselves are not implanted. In the case of Neurocam, algorithms have replaced the Cutters and the memory workers of Rekall Corporation, ostensibly augmenting our brain's ability to remember and respond to sensory data. Neurogeographies serve as repositories for individual and collective memories; they may be radical, re-mapping space and place, but they are also becoming essential tools in the outsourced world of neurotechnologies that include EEG wearables.

Navigating the "Neuropolis"

Neurogeographies are becoming ubiquitous ways to interact with built environments. Growing from roots in psychogeographies and bio-mapping, they have evolved into entrepreneurial projects using EEG for tracking and mapping arousal. Neurogeographies are not entirely novel, but they do introduce and rely on a host of data-driven EEG wearables that allow for different aggregations of data. How the data are represented, valued, and evaluated will help determine the usefulness of neurogeographies for individual and collective decision making.

Much of the work on neurogeographies remains speculative—like the plight of Steve and his smartphone. Yet, as Tom Stafford has argued, the potential for our wired future is immense—and potentially different from the types of revolutions we have encountered thus far. "Bio Mapping promises to open up the field of possibilities for social hacking, for awareness hacking, for a rediscovery of conscious collective action—not along the lines of the monolithic political movements of the 20th century, but in swarms, collectives and communities. The hacker ethic is one with the possibility to combine individuality with interconnectedness" (Stafford 2009, 94–95). We might also consider neurogeography in tandem with what Des Fitzgerald and colleagues have termed "the neuropolis," which they characterize as "the city understood as a matrix of transactions between urban life and the always-developing, malleable brains of urban citizens" (Fitzgerald, Rose, and Singh 2016, 223). For these researchers, the concern is the effects of urban stress on the mental health of a city's inhabitants. In the case of MindRider, too, the mapping of anxiety is central to reconsiderations of the city and its infrastructures. Neurogeographers of the future surely share this hope for a more radical hacking of cityscapes. The MindRiders and the Neuro Taggers each seeking some combination of individual awareness and collective cartographic change. These are walkers (and riders) with a thesis, even and especially if the outcomes for neurogeographies are not yet established.

From Soufflé to Signs of Death

Instrumental Intimacy about Us, without Us

In 2015, a major restaurant chain in Japan offered up its workers and its new menu item, a soufflé, to help associate professor Yasue Mitsukura test her EEG-based brain wave analysis system. Her instrument, like many of the EEG wearables we have encountered in these pages, is intended to measure specific states of arousal—in this case, "approval, interest, concentration, stress and sleepiness." Participants were instructed to open their eyes only after the soufflé was served; at that time, the machine began to record electrical activity. On the whole, subjects registered electrical activity levels that seemed to indicate enjoyment in terms of presentation and taste. But then the participants attempted to scoop up the fruit topping: one woman's " 'stress' level soared to 90. She admitted she didn't know where to put her spoon, and felt the dessert was difficult to eat": "the tests revealed it was difficult to scoop up the fruit with the small spoon provided. Now managers with the restaurant chain are considering serving the soufflé with a fork instead. It is an adjustment no one would have thought of without the testing. 'The system allows us to measure emotional changes in minute detail, and in real time,' says Yutaka Ogawa, an executive with the restaurant chain" ("Brainwaves for Business" 2015). The news story represents the outcome of this experiment (a simple change from spoon to fork) in a couple of intriguing ways: first, it shifts the terms of what is being measured from arousal to "emotion." Second, the possibility of switching from spoon to fork is not merely a quick fix but is instead "an adjustment no one would have thought of without the testing." In this telling, our interaction with a plate of food demands EEG-based intervention: knowing how to adjust for issues of arousal requires an instrumental guide.

In one respect, the soufflé is another example of the ways that instrumental intimacy—knowing and changing ourselves through our interactions with various technologies—creates solutions to the problems it has simultaneously rendered visible, measurable, and changeable; we can know ourselves better and therefore more easily intervene in our behaviors. But the soufflé also demonstrates another phenomenon shared by EEG wearables: the fact that interested parties may also take advantage of users through the electrical information collected from their brains. In this restaurant experiment, public displays of arousal are instrumentally useful to others. In short, instrumental intimacy can be used to gather information about us and make decisions without us in both hilarious and utterly serious ways. From soufflés to signs of death, I analyze some of the implications of making visible the electrical activity of our brains.

As I indicated in chapter 2, the shifting characterizations of brain activity from unconscious to conscious (and sometimes back again) allow for claims that EEG wearables reveal things that users either do not know or cannot consciously control, thus leaving open the possibility that companies could use data gathered from users to better market their products. To complicate matters, EEG wearables are not yet regulated; the massive amounts of data produced by these technologies are often collected by the parent companies who can use them for undisclosed purposes. Take, for example, a recent study about consumer preference in which brain activity is discursively—and conveniently—rendered unconscious and unavailable to us: "The general assumption is that human brain activity can provide marketers with information not obtainable via conventional marketing research methods (e.g., interviews, questionnaires, focus groups). . . . This is mainly driven by the fact that people cannot (or do not want to) fully explain their preferences when explicitly asked; as human behavior can be (and is) driven by processes operating below the level of conscious awareness" (Khushaba et al. 2013, 3803). Herein lies the paradox succinctly stated: EEG can be used to open access to the once inaccessible and to create access to the still inaccessible. Divisions between and applications of the two types of access are crafted by wearable manufacturers and researchers alike. The user is not always constructed as the active beneficiary. In this scenario, EEG is expected to work like lie detection equipment: electrical brain activation is assumed to be out of our conscious control, but whoever can map it can make use of it.

In 1972, as Joe Kamiya's experiments on brain training were garnering popular media attention, the Mind Institute was offering to train the businessman's brain, and new states of mind were being minted, other EEG experiments were under way to test employees for desirable personality traits. Ray Cooper and colleagues, working at the Burden Neurological Institute, gained attention for their experiments concerning various evoked brain potentials. While Cooper was unwilling to speculate about the potential instrumental use of the institute's studies, *Nation's Business* news was already making all of the speculative connections. The headline: "Careers Could Ride on Brain Waves"; the takeaway for business executives making decisions about potential employees:

> One day, not too many years from now, you may have an electroencephalograph made on a candidate for a highly ranked job with your company. And from that "EEG" you may be able to find out if he:
>
> - Is a decision-maker
> - Is enough of a risk-taker, yet not too aggressive, to seize opportunities to push ahead with the company
> - Is a "visual" or "abstract" thinker
> - Has too much, or too little, anxiety
> - Can anticipate problems
> - Reacts quickly to changing situations and altered orders
> - Has adequate powers of concentration (1972, 58)

In this nearly exhaustive Goldilocks list of "not too cold and not too hot," the business newspaper synthesizes and converts Cooper and colleagues' work on reaction time, sequence recognition, concentration, and other factors into desirable employee characteristics—characteristics that coincide with a system of mental disposition affiliated with capitalistic business practices. Note that the discourses of EEG are not about training and empowerment for the wearer but diagnostic advantages for the employer. Although Cooper is not willing to make blanket statements about translating EEG data into actionable policy, and he is quoted as saying, "Frankly, I wouldn't want this business with EEG to go too far" (61), he also speculates about the potential use value of the institute's work: "Certain jobs call for more obsessiveness than others. Computer programmers must be obsessed with getting perfect information into their machines. Retailers must be somewhat obsessive in getting things sold and having them displayed perfectly" (60). In this application of EEG to job performance and the potential to monitor employee brain states, we might even hear echoes of the SmartCap. And although "fatigue" might seem like less of a translation than "obsession," similar logics are at play.

Cooper also discusses both control and comparative training. One of his main concerns is bringing brain activity under control: "Obsessiveness is cousin of determination if it is under control. A person's obsessiveness isn't under control if he pursues a failing policy right out of the window." It is one thing to possess the raw brain activation, quite another to harness that activation into desirable personality traits. "We want to categorize people we test. I wish we could get 50 top executives who have done well in business, take their EEG's and then take the EEG's of 50 businessmen who didn't make it quite so well. I would like to compare the EEG's of the two groups. I think we eventually have to do something like that" ("Careers" 1972, 61). In his speculative hopes, Cooper predicts some of the bases of accelerated learning that Chris Berka has since taken up in Advanced Brain Monitoring: find the experts, map their electrical activity, and then establish ways to compare other, novice—or less successful players—to them. Berka's studies run to both ends of the instrumental spectrum: although the training is situated in the subject's hands—or brain, as the case may be—funding for at least some of the accelerated learning studies came from DARPA and, one would suspect, would therefore be used for more instrumental ends.

Perhaps the ultimate example of instrumental intimacy about us without us is the advent of "brain death" as a concept and diagnostic tool. Now often confirmed through a particular sequence of EEG evaluations, the invention of "brain death" was not a simple matter and stirred many ethical controversies in the decades leading up to its acceptance by medical communities around the world.[1] I do not belabor those histories here, as other scholars have done a fine job of working through the ethical dilemmas and the cases for acceptance. Instead, let us look at some of the ways that brain death is an excellent—and ultimate—example of instrumental intimacy.

As with other states of mind, including REM sleep and conscious brain training, brain death (BD) in humans became a measurable state of mind only after the advent of EEG. As Calixto Machado and colleagues argue in the *Journal of Medical Ethics*, "the BD concept was supported by the discovery of EEG." They detail the basic sequence of events as follows: "In 1929, Berger first discovers and records EEG in human beings, which he termed 'elektroenkephalogramm.' Within a year, Crue et al. attributed 'the cause of clinical death to the fall in the potential between the different tissues,' and 'the cause of the death of single cells or of tissue cultures to the fall in the potential on the cell membrane'" (Machado et al. 2007, 197). Despite these initial forays into brains and death, the idea of brain death as a state of mind did not come to fruition for another forty years.

Between 1963 and 1968, EEG was championed by several doctors and committees as the best measure for the new state of mind, even and in spite of a battery of neurological tests that can help to determine unresponsiveness without the help of machines.[2] Hannibal Hamlin presented some of the first arguments at the American Medical Association in 1964. Buried in the back pages of the *Chicago Tribune*, one headline reads, "Brain Signals Death of Man, Doctor Says." The copy goes into some detail: "A machine which measures electrical currents generated by the brain should be the first arbiter to tell doctors when a dying patient has reached the 'point of no return,' . . . Dr. Hannibal Hamlin of Harvard University Medical School said the device, called an electroencephalograph, is much more reliable than listening for fading heart sounds to determine when death has become irreversible" (Gibbons 1964, 25). Although numerous issues would be raised in the next five years, the rise of hospital deaths, the advent of intensive care, and questions of organ transplantation helped to foist EEG into the limelight as the next best way to determine if a patient is really deceased. By 1969, newspaper advice columns—including Dr. Van Dellen's (see chapter 3)—had largely accepted EEG: "Deciding whether the individual is alive or dead is not difficult if an electroencephalograph is available to check the electrical waves of the brain" (1969, 16). This popular column's appraisal came just one year after the Sydney Declaration concerning human death and a Harvard committee report on the acceptance of brain death as a primary criterion for bodily death: "The Harvard Report had momentous repercussions and constituted a breakthrough account, establishing a paradigm for defining death by neurological criteria" (Machado et al. 2007, 198). By 1975, even the medical ethics journals were publishing essays by neurosurgeons who not only supported the technique but expended much ink countering arguments against EEG as a measure of brain death, noting "the diagnosis of brain death can be made confidently, as is already common practice, and this should become standard good medical practice" (Jennett 1975, 63). While questions remained for some lay audiences, reports in popular news columns and in medical journals helped to strengthen the hold of EEG and its ability to recognize a final state of mind.

In the case of EEG wearables, instrumental intimacy is about using machines to reveal otherwise inaccessible details about our physiology—and therefore our states of mind. But because ideologies of instrumental intimacy create access points, those points can also be used to make decisions about us, without us. Diagnoses of brain death, in particular, construct a machinist interface by which to gather data about bodies—and brains—that are no longer communicative and may be beyond efforts at arousal. If the other brain states discussed in chapters 2 and 3 depended on finding arousal where there was once assumed to be stasis, then brain death is an absence of arousal, but this distinction only becomes possible in a post-EEG cultural context in which arousal is assumed to be the norm for a conscious, and therefore living, brain.

In the intervening decades, machines have become one of the primary interfaces for determining death. And if we are concerned, as were laypeople and medical experts alike, that the machines might miss something or that our perceptions of a "flat" EEG could be affected by complicating and unknown factors such as drug consumption, the solution sometimes proposed is simply more machines. In 1969, as brain death was coming to be an accepted state of mind, concerns about user and machine error were answered in this way: "Dr. Bickford said that computers are being programmed to read EEGs and make the determination. This, he said, will be far more accurate than having personnel 'eyeball' the EEG readings"; in addition, new standards of measurement were introduced: "EEG personnel are advised to use enough electrodes to make sure the whole brain is dead because it is possible for focal points of life to exist. They are also advised to take readings at high magnification to make sure that there is not electrical activity at lower levels. They are advised to observe the flat EEG for at least 30 minutes and to repeat the procedure 24 hours later" (H. Nelson 1969, 3, 25). In both cases, the answer is not to interact with the human, to touch, or to listen, nor is it to reconfigure our definition of death so that is it broader or more complex than the absence of electricity. Instead, death becomes another state of mind to be visualized, measured, and determined through instrumental intimacy.

Throughout this book's case studies, I have illustrated how EEG, a largely overlooked neuroscientific technology has left the confines of the laboratory to become a public, fashionable means to achieve better brain control. Print news media sources, advertisements and company copy, scientific studies, and user-generated data illuminate the discourses of control, responsibility, and risk that inform EEG wearables emerging in the twenty-first century. And in this final coda to the book, I revealed some of the ways that the rise of instrumental intimacy has been, and will continue to be, used to reveal things about us, without us. EEG wearables are one among many machines that could guide us in this century; whether we follow as wearers or watchers, we would do well to remember that we may not need a machine to tell us if we like the soufflé.

Notes

INTRODUCTION: A "Machine to Guide Them"

1. Perhaps this omission is because EEG is simply less sexy than its brain-imaging counterparts. For nearly a century, EEG has required a cumbersome, wet, laboratory preparation involving an electrode cap and hundreds of training wires; its output, an oscillographic representation of the brain's electrical activity, does not offer provocative visualizations for the popular media to publicize. Or perhaps EEG has escaped sustained criticism because, as a technology, it has been relatively unscathed by the neuroscientific controversies that populated the early-twenty-first century: with the exception of Lawrence Farwell's application of EEG to a technique he called "brain fingerprinting," EEG has stood the test of time as a stalwart, laboratory-based technique whose controversies have been largely relegated to its past. Indeed, historians of science have spilled much ink over Hans Berger's initial electrical experiments in the 1920s, his radical ideas about the materiality of thought, and the long and troubled narrative leading to EEG's eventual acceptance, but EEG has gone relatively unnoticed in contemporary debates that led to the rise of what's been called *critical neuroscience* and the disciplinary appeals of the neuroscientific turn.

Excellent studies of individual neuroscientific and other imaging technologies include: Joseph Dumit's *Picturing Personhood* (2004), Kelly Joyce's *Magnetic Appeal* (2008), and Barry F. Saunders's *CT Suite* (2008). For more information about critical neuroscience, see Choudhury and Slaby (2011).

2. Only a century earlier—and just before Hans Berger began experimenting on the human brain's electrical potential—Hugo Munsterberg characterized his psychophysiological instrument of choice, the chronoscope, as a "mental microscope" because "it makes visible that which remains otherwise invisible, and shows minute facts which allow a clear diagnosis" (1908, 77–78).

3. "The metaphor of the map," notes Anne Beaulieu in her study of brain-bank databases, "helps concretise what are highly abstract manipulations of data into an endeavour of surveying and exploration. They help make tangible the virtual object of research as a totalising exploration of a territory" (2004, 378).

There are many atlases of electroencephalography, and many have grown more and more specific with each passing decade. One of the oldest is Frederic and Erna L. Gibbs's *Atlas of Electroencephalography* (1941).

4. I should note here that there is one important difference between NeuroVigil and these other, direct-to-consumer companies: NeuroVigil only markets its products to a select group of researchers, medical experts, and government agencies. It does not sell iBrain directly to consumers.

5. Many of these market characteristics are anticipated by Gilles Deleuze's portrait of a post-disciplinary social framework in "Postscript on the Societies of Control" (1992).

6. The study in question is from Charles Grau and colleagues out of Starlab in Barcelona. The team's experiment claims that they were able to successfully elicit brain-to-brain (B2B) communication by transmitting electrical impulses from one brain to another via EEG and the Internet (Grau et al. 2014).

7. The *Oxford English Dictionary* contains two definitions for *brain wave*: the first is figured in terms of telepathy and dates to 1869. So far as I have been able to trace, James Thomas Knowles coined the term *brain-wave* in a letter to the editor of the *Spectator* (1869). The editor, who leaves his response unsigned, summarily dismisses "J.T.K.'s" theory in a column printed in the same issue ("The Hypothesis of Brain-Waves" 1869). That a term such as *brain-waves* could emerge in the late nineteenth century makes sense given a larger cultural and scientific focus on metaphysics, the rise of the telegraph in the 1830s and 1840s (see Otis 2001 for an excellent analysis of connections between the two), and theories of thermodynamics and the ether (Clarke and Henderson 2002; Littlefield 2011). But these early theories of "brain-waves" are largely focused on potential communication or telepathy (Luckhurst 2002). If we look to references in fiction from the period, "brain waves" are often shorthand for the spark of an idea or invention (Bonner 1890; Ridge 1917). For an interesting meta-analysis based on n-grams, see Demarest (2016). The second definition of *brain wave* in the *OED* references an article in *Discovery: A Monthly Popular Journal of Knowledge* from 1935. In this post-EEG context, the authors move away from thought-action (which was closer to J. T. Knowles's usage) and toward a physiological definition: "By means of electrical records made through the skull various states of the brain can be recognized; but the 'brain waves' thus far recorded do not appear to be the result of thought-action" ("Notes of the Month" 1935, 1).

8. In one of his many historical analyses of EEG, Cornelius Borck aptly argues that "before their observation, brain waves did not yet exist as something awaiting discovery" (2008, 370). However, I would qualify his remark based on the earlier occurrences of the phrase *brain-waves* in the popular media of the late nineteenth and early twentieth centuries.

9. Earlier experiments by Richard Caton (1875)—based on work by Luigi Galvani (1791)—Adolf Beck (1891), Vladimir Neminsky (1913), and Napoleon Cybulski (1914) all established that there was electrical activity in the body and brain of animals. See Ahmed and Cash's 2013 review article on desynchronized EEG for a more complete history of electrical experimentation and the brain. Hans Berger claims to have identified the human electroencephalogram in a 1924 experiment; his first publication in the topic was printed in 1929. However, Berger's experiments were not validated until the 1930s—thus the time scale of the 1920s to the 1930s.

10. By "noninvasive," I mean that no open-skull surgery is required for current EEG. When Berger was first devising the EEG, he did experiment on patients who were undergoing surgery for other reasons, and, as Herbert Jasper notes in his preface to Gloor's translation of Berger's papers, Berger made "use of the many patients with skull defects resulting from [World War I] to prove that the electrical waves recorded from the intact skull were truly representative of the electrical activity of the brain beneath" (Gloor 1969, vi). In his noninvasive work, Berger used implanted needle electrodes. Adrian and Matthews used "lint covered squares of copper gauze soaked in saline" (1934, 3).

11. Pierre Gloor (1969) provides translations of all fourteen papers from Berger's original experiments.

12. Originally, Berger employed an "Edelmann string galvanometer, an instrument which had been designed to record electrocardiograms." He went on to use a double-coil

Seimens galvanometer, which was likewise intended for electrocardiograms; only in 1932 did the Siemens Company construct an oscillograph for him that recorded voltage rather than current (Gloor 1969, 7).

13. From here onward, I am quoting from Peter Gloor's excellent translation of Berger's papers (Gloor 1969).

14. While scientists have now confirmed these potentials and added three more to EEG recording, the alpha rhythm was often referred to simply as the Berger rhythm.

15. According to Cornelius Borck's account, E. D. Adrian and B. H. C. Matthews heard about Berger's experiments thanks to a group working at the Berlin-Buch Institute in 1932 (2001, 580).

16. As Peter Gloor, David Millett, and Cornelius Borck argue based on evidence from Berger's diaries, Berger's initial inspiration for attempting to trace the electrical activity of the brain came from a frightful near accident and its repercussions in the form of familial communications. After he was nearly run over by a wheeled gun during a morning military exercise, Berger received a telegram from his father (at his sister's insistence) asking about his well-being. According to Berger's diary, he believed this was a "case of spontaneous telepathy in which at a time of mortal danger, and as I contemplated certain death, I transmitted my thoughts, while my sister, who was particularly close to me, acted as receiver." Gloor notes, "Berger's decision to abandon the study of astronomy and to devote himself to the elucidation of the relationships between the psychical and the physical world was prompted by this curious event which convinced him of the reality of telepathy" (1969, 3). And to this study of the materiality of psychic events, Berger devoted his life—largely in secret.

Berger tried many theories before finally settling on electrical measurements of the scalp. David Millett argues that "understanding [Hans] Berger's path to the EEG within the context of his own scientific era and life experiences . . . provides a historical prelude to the modern investigation of the brain and its functions" (2001, 522). Perhaps the most complex (and somewhat convoluted) was his conception of psychical energy. As Gloor explains, "Since the only source of energy in the brain is chemical, the problem from a conceptual point of view appeared simple: in the brain chemical energy is transformed into heat, electrical energy and psychical energy" (1969, 15). This equation led to studies on brain temperature, which ultimately failed, but it—like other aspects of Berger's work—was a product of his time, a time when the theory of thermodynamics was all the rage and efforts at making thought material could be found in the science and science fiction of the age (Clarke and Henderson 2004; Littlefield 2011; Millett 2001; Sourkes 2006).

In his diary, Berger spoke of a *Hirnspiegel*, which has been translated as "cerebroscope" and, more literally, "brain mirror" (Millett 2001, 535)—on the one hand, this term captures the materiality of Berger's interests, but on the other it illuminates what Berger may have been theoretically and philosophically imagining for his technology, hopes that have somehow come to a weirdly actualized fruition with the twenty-first-century's spate of EEG wearables. Recent articles concerning brain-computer interfaces also indicate the continued hope that EEG could be used for alternate brain-to-brain communication (Wolpaw et al. 2002, 768). NeuroVigil's iBrain represents new possibilities for individuals who wish to use their "thoughts as signals"; the company has tested its device, most famously, with Stephen Hawking (Duncan 2012).

17. Social histories of Hans Berger and his EEG sometimes linger on the troubles Berger underwent en route to his great discovery or speculate about why he went

unrecognized for so long: some credit Berger's personal affect, which is described as "shy, reticent, and inhibited" (Ginzberg 1949, 370), or the fact that he worked in secret and only in the evenings after the work in his Jena clinic was complete; others cite the Nazi culture of Weimar Germany and Berger's resistance to fascism (Gloor 1969; Borck 2001).

18. Jonna Brenninkmeijer remarks that Berger's liminal scientific position was the result of numerous factors. First, psychophysiology "had fallen into 'disrepute among neurologists and psychiatrists' (Gloor, 1969, p. 3) at the end of the century." Second, Berger did not follow either of the more popular schools of thought: the neuroanatomical approach or the functional approach. Finally, Berger's focus on recording the electrical activity of the human brain was regarded as irrelevant and useless: "Most scientists in that period . . . simply did not believe in electrical measurements of the brain and considered the (weak) electrical oscillations as artefacts of the apparatuses." In short, Berger chose a novel path, drawing "his inspiration from the electrophysiological experiments on animals from Caton, Fleisch von Marxow, Beck and Cybulski" (Brenninkmeijer 2013b, 111).

19. Examples of EEG as "the brain machine" can be found in a 1955 *Flash Gordon* television episode called "The Brain Machine" and a 1955 British thriller of the same title.

20. The complexity of EEG should be noted here: "As the electrical potentials generated by neurons have to travel a relatively long distance through the surrounding tissue to the scalp surface and due to sheer number of neurons within different brain regions, surface EEG cannot record the activity of individual neurons. Rather, it represents the summation of electrical activity of thousands to millions of spatially aligned neurons" (Mihajlović et al. 2015, 8).

21. I would stress, as does Wexler (2015, 686), that the FDA's policies concerning the regulation of various devices can and have changed over time. Future regulations may focus on issues of physical risk and issues of data/information privacy. Among science and technology studies (STS) and legal scholars, debates about regulation have focused on cognitive enhancement devices or DIY neuroscientific stimulation devices such as transcranial direct current stimulation (TDCS). For further reading, see Kostyukovsky (2015), Maslen et al. (2014, 2015), Piwek et al. (2016), and Wexler (2015). For information on other countries' regulatory policies, see Ou and Liu (2016) and Sung (2015). For accounts in the popular press, see Lecher (2015), for example.

22. Historians of science and STS scholars have discussed the rise of inscription at some length. Commentators have addressed the work of Charles Féré, Christian Ruckmick, William Marston, Maurice Schiff, Angelo Mosso, Alfred Lehmann, and Alfred Binet, among others.

23. Several emergent STS scholars are working to bring affect theory and technology together in very intriguing ways. The work of Kelly Underman, for example, seeks to explain the management of emotion within and between practitioners and patients as forms of "affective practice" that "reshape the body's capacity to feel" so that emotional management becomes part of a new natural-seeming habit (2015, 180). Mariana Craciun's work on psychotherapy likewise theorizes the problem of practitioner emotions that become embodied as "affective-relational" expertise—tools that can be part of a larger, quantified therapeutics (2013, 4).

24. The parenthetical hesitation alludes to the fact that tracking ourselves is ostensibly a choice but also a responsibility. We do not have to be actively invested in self-tracking to be willing to participate in the regimes of control that demand records of modulation.

25. For a broader perspective on aging and its complex relationships to technology, see Joyce and Loe (2016).

26. If you are interested in seeing Cartwright's work, his website, www.stephencart-wright.com, details many of the self-tracking projects he has been engaged in since the late 1990s.

27. Martyn Pickersgill and colleagues have been working on questions about how information about the brain is being consumed by various publics: "When the brain is consumed, who is controlling these products? How are they negotiated by consumers? And how do 'users' accept, reconfigure and resist the 'ontological imaginaries' . . . that may be embedded in or produced by these services, products and cultural artefacts?" (Pickersgill, Cunningham-Burley and Martin 2011, 362). As I demonstrate in chapter 4, users are inventing new ways to interact with and draw information from EEG wearables; however, these uses remain inflected by discourses of deficiency, visualization, and control.

CHAPTER ONE: Public Displays of Arousal

1. In the introduction, I included Berger's own origin story of the EEG. According to David Millett's interpretation of the same tale, "There seemed to be no escaping the conclusion that Berger's intense feelings of terror had assumed a physical form and reached his sister several hundred miles away—in other words, Berger and his sister had communicated by mental telepathy. Berger never forgot this experience, and it marked the starting point of a life-long career in psychophysics" (2001, 524).

2. Although NEUROTiQ is not currently on the market, Sensoree does sell an emotion-displaying Mood Sweater that "lights up different colors depending on your mood" (Kunz 2014).

3. In his brief news report about the device, Eddie Krassenstein (2014) characterizes the "fabric" as follows: "To make the headdress both comfortable and provide it with more ability to move, Neidlinger combined the technology of 3D printing with that of traditional knitting. The globules were then embedded along with some electronics within a nylon knit, that took approximately 61 hours to create. The 3D printing of the globules was done on a Form 1 3D printer, and took 8–10 hours each to print out. In total, the 3D printing process took 102 hours."

4. Neidlinger is not alone in her refashioning of bodily communication. Recently, a group of collaborators designed and executed a Neuro Knitting project that knitted a user's brain wave activity into a fabric (Guljajeva, Canet, and Mealla 2012).

5. In his larger project, Malin discusses the cultures in which psychophysiological technologies—particularly those focused on translating *emotion* from bodies—were developed and deployed. The cases he examines from the 1930s reveal an American culture obsessed with emotional control and maintenance in the face of mass market and technological overstimulation.

6. For more on biomimicry see Lakhtakia and Martin-Palma (2013) and Benyus (1997), among other, more disciplinary-specific texts.

7. Before the SmartCap, several other fatigue-monitoring systems were implemented to limited success. These largely involved in-cab cameras intended to measure "3-D head and facial movements, PERCLOS (percentage eyelid-closure), and micro-sleep events that indicate signs of fatigue or tiredness" (GuardVant 2015). Also see the nap zapper, www.amazon.com/Generic-Nap-Zapper-Anti-Sleep-Alarm/dp/B000BK4KW8.

8. As I mentioned earlier, the SmartCap is also available as a woolen hat and a headband. It is not yet available in a hardhat.

9. Gardner and Wray (2013) provide a helpful analysis of the marketing strategies of the parent companies of EEG equipment I cover in this book, including NeuroSky and EPOC. In addition, their analysis covers questions of gender, particularly in EEG wearable advertisements.

10. VaenMousavi et al. (2007) rely on older studies for their division between the terms, namely Barry et al. (2005) and Pribram and McGuiness (1975, 1992). See also related studies intended to distinguish between arousal and activation, such as Barry et al. 2007.

11. The caveat here—and it is an interesting one—is that NEUROTiQ displays are somewhat opaque. Without a color scale in front of you, the emotive headdress only tells you that the electrodes are registering activity of one kind or another.

12. In name alone, NEUROTiQ evokes a century of Freudian psychology, including diagnostic categories such as the "neurotic." Neidlinger's moniker for her headdress plays with this designation and with the combination of "neuro" and "tiq"—a slang word for "boutique," long associated with fashion. Whether or not Neidlinger is aware of her reference (she has not spoken of Lacan's use of "extimacy," for example, in any of the press I have read or listened to), what matters is the cultural resonance this term evokes for users who may see it as a play on definitions of neuroticism.

13. The company has also developed a smartphone app that visualizes the data for user and supervisor.

CHAPTER TWO: In the Zone

1. For example, an individual may wish to produce more alpha waves, which have been linked to relaxation and focus; their training would include a sensory feedback signal when their brain is producing alpha waves. The subject would then try to maintain the haptic signal (a pulse of light, a particular sound) as a means of continually producing the desired, alpha electrical activity. On the distinction between arousal and activation, see chapter 1.

2. The institute, which has locations in Canada, the United States, and Germany, is the brainchild of Dr. James V. Hardt, who holds a doctoral degree in Psychology from Carnegie-Mellon University.

3. The Biocybernaut Institute may be the most inaccessible of the neurofeedback options available, given the cost ($10,000–$20,000 for a five-to-seven-day session); however, the basic tenets of its operation mirror those of elite athlete training facilities as well as lower-end products such as the Lotus or Puzzlebox Bloom, which are EEG neurofeedback trainers intended for home use and costing about $150. At the Biocybernaut Institute, the "scientifically measured benefits" for the seven-day treatment include "a 50% increase in creativity and an 11.7 point increase in IQ. In addition, there are big increases in motivation, mental clarity, ability to focus, and reductions in anxiety and depression, as well as improvement in mental and emotional well-being" (Biocybernaut Institute 2016). Whether some of these promises seem farfetched matters less than the fact that they are metrics, offering quantification of neurofeedback training. As this chapter demonstrates, systems of quantification are products and producers of the shift in assumptions about conscious control and mastery.

4. Deficit models precipitate discussions of enhancement, which I discuss in the next section on mastery below.

5. Neurofeedback has a long and storied (some would argue dubious) history that has been recounted by numerous scholars, journalists, and pundits (Brenninkmeijer 2010, 2013a, 2013b, 2016; Linden 2014; Robbins 2008). In his recent book, *Brain Control: Developments in*

Theory and Implications for Society, David Linden traces the seeds of neurofeedback into the nineteenth century, but he argues that the first neurofeedback experiment took place in the lab of Adrian and Mathews in 1934 as they tested the Berger Rhythm (2014, 14, 9). Although these two scientists engaged in what we would now term *neurofeedback*, they did not use the term at the time. Indeed, the term *neurofeedback* emerged in the 1960s, thanks to Joe Kamiya's article in *Psychology Today*. Preceding Kamiya's work was the research of William Grey Walter in the 1950s and Barry Sternman's experiments with cats in the 1960s.

Professional societies include "the Association for Applied Psychophysiology and Biofeedback (AAPB); the International Society for Neurofeedback and Research (ISNR), and the Biofeedback Association (BFE) [which] promote the clinical use of these technologies and certify their practitioners" (Linden 2014, 106–7).

6. As an aside and a corollary case, it is worth noting J. H. Blair's experiments published as "Development of Voluntary Control" (1901), which involved the movement of ear muscles.

7. In her dissertation, Brenninkmeijer interviews people undergoing neurofeedback who describe the process in terms of reprogramming a machine: "They describe the neurofeedback process in terms of 'a defragmentation of your computer,' 'cubes put in the right order,' 'a computer wiring me,' 'a re-programming of my brain,' 'my system is unstable,' 'my systems resets itself over and over again'" (2013b, 64).

8. The couple's story is covered in great depth in Jim Robbins's book *A Symphony in the Brain*. The Othmers' son, Brian, was an important link in corporate development of neurofeedback.

9. Sports psychologists have long trained athletes interested in improving the so-called mental game, certain aspects of which overlap with neurofeedback training. The main differences here are the technologies and the biomarkers used for determining how much progress is being made.

10. There are different kinds of mental gymnasia. Some, which may or may not include EEG technologies for neurofeedback, incorporate physical training with various types of neuro-training. Still other examples include Neurobics Clubs, marketed to aging baby boomers. These neuro-aerobics spaces are typically filled with computers and gaming software intended to challenge participants to stay sharp and exercise their mental acuity, but they do not usually include neurofeedback equipment or protocols.

11. The Mindroom approach was "originally developed for Italy's AC Milan soccer team by Professor Bruno Demichelis, founder and former Scientific Director of Milan Lab" (Mindroom 2015a).

12. Thought Technology's involvement in sport goes back over three decades. Mind over Muscle, the Mental Gym, was co-created by Major Nory Laderoute, former director of physical education training for the Canadian Armed Forces Combat Training Center and Thought Technology in 1974, and Lawrence Klein, vice president of Thought Technology. It teaches athletes relaxation and visualization to enhance sports performance and has been used by thousands of Olympic athletes worldwide (Mindroom 2015c).

CHAPTER THREE: "Such a Natural Thing"

1. Take, for example, the following claim: "Our findings indicate that sleep in industrial societies has not been reduced below a level that is normal for most of our species' evolutionary history" (Yetish et al. 2015, 2867).

2. See, for example, Brunswick (1924).

3. For more information on the history of sleep science and sleep medicine, see Morrison (2013) and Shepard et al. (2005). See also personal accounts by Dement (1990) and Foulkes (1996). See also Gottesmann (2001, 2005), on whom Morrison relies: "From 1953 through 1964, 181 articles appeared in the human and animal sleep literature. During the following two years, the subject of an entire book, *The Golden Age of Rapid Eye Movement Sleep Discoveries: 1965–1966*, at least 325 were published" (Morrison 2013, 399).

4. There are varying degrees of acceptance as to whether Kleitman was interested in making connections between REM and dreaming; Aserinsky is said to have been mildly interested (and indeed, their paper in *Science* does draw this connection), but the primary proponent of the REM sleep-dream connection is Dement.

5. Foulkes provides an excellent list of the "major laboratories participating in this research in the 1960s and early 1970s," including "CUNY (Antrobus, Arkin), Montefiore (Herman, Roffwarg, Ellman), Downstate (Shapiro, Witkin, Goodenough), Mt. Sinai Hospital (Fisher, Kahn), Maimonides (Ullman, Krippner), Boston VA (Greenberg, Hartmann), Cincinnati (Kramer), Chicago (Rechtschaffen, Vogel), Illinois at Chicago (Cartwright, Monroe), Virginia (Van de Castle, Hauri), Stanford and UC Davis (Tart), Texas (Cohen), Wyoming (Foulkes), and Oregon (Breger). In addition, persons later to figure in the development of dream research gained experience in laboratories devoted more specifically to sleep (e.g., Lavie at Webb's laboratory in Florida and Kripke's in San Diego)" (1996, 614).

6. See Browning (1956), Franke (1958), and Robinson (1959). The section epigraph comes from "Psychologists Get Some New Ideas" (1960).

7. One exception comes tangentially, from a review of Edwin Diamond's 1962 book *The Science of Dreams*. In it, reviewer Irwin Kremin casts doubt on the mechanistic advances (such as EEG) given that sleep researchers must still wake subjects and ask them to scribble down their dream memories. He chastises Diamond for championing the technologies ahead of their actual use: "Mr. Diamond apparently also believes that these dream reports, by grace of the new electronic techniques, are somehow more real, more valid, more detailed, than was otherwise so. What he fails to realize is that electronic equipment does not of itself suffice to make matters neatly precise and objective" (Kremin 1962).

CHAPTER FOUR: Neurogeography and the City

1. As Sinclair puts it, "this was walking with a thesis. With a prey. . . . The stalker is a stroller who sweats, a stroller who knows where he is going, but not why or how" (Coverly 2006, 120).

2. See, for example, Coverly (2006).

3. For a discussion of current definitions of, and issues with, arousal in neuroscience, see chapter 1.

4. I use the hybrid *subjective-objective* here in an attempt to capture Ducao and colleagues' sense that their data are subjective (Ducao 2014, 57) even though they are recorded by a graphic inscription device (EEG) that many still consider to be objective.

CONCLUSION: From Soufflé to Signs of Death

1. Take, for example, issues of death at home versus death in the hospital—particularly the rise of intensive care and artificial respirations, as well as questions of coma and vegetative states; of death and organ transplantation timing (Machado et al. 2007); of availability and expertise (Fermaglich 1971); and of acceptance within the medical community (Jennett 1975, 63).

2. As Machado et al. note in their 2007 article, precursors include Wertheimer, Jouvet, and Descots (1959) and Jouvet (1959), who described the "death of the nervous system"; that same year, Mollaret and Goulon (1959) discussed deep coma in terms of EEG, but even "they did not consider their patient to be dead" (Machado et al. 2007, 198); finally, in 1963 Alexandre used a brain death diagnosis (via EEG), which was similar to that of the Harvard Committee, as justification for the first organ transplant. For a more complete history, see Machado et al. 2007, 198–99.

References

"Accelerated Learning: How to Get Good at Anything in 20 Hours." 2013. YouTube video, 23:27. Posted by Good Life Project, June 26. www.youtube.com/watch?v=IB6K6 omkmho.

Adrian, E. D., and B. H. C. Matthews. 1934. "The Berger Rhythm: Potential Changes from the Occipital Lobes in Man." *Brain* 57: 355–85. Republished in *Brain* 133 (2010): 3–6.

Ahmed, Omar, and Sydney Cash. 2013. "Finding Synchrony in the Desynchronized EEG: The History and Interpretations of Gamma Rhythms." *Frontiers in Integrative Neuroscience* 7: 58. doi:10.3389/fnint.2013.00058.

Alessi, Anthony. 2012. "Training the Brain for High Speed Decision Making." *Healthy Sports* (blog), July. http://dralessi.blogspot.com/2012/07/training-brain-for-high-speed -decision.html.

Angier, Natalie. 1995. "Modern Life Suppresses an Ancient Body Rhythm." *New York Times*, March 14. www.nytimes.com/1995/03/14/science/modern-life-suppresses-an-ancient -body-rhythm.html.

"artfuture:: Maker Faire 2010—Kristin Neidlinger—GER." 2011. YouTube video, 4:00. Posted by artfuture, June 15. www.youtube.com/watch?v=QIuobvbvRso.

Aserinsky, Eugene, and Nathaniel Kleitman. 1953. "Regularly Occuring Periods of Eye Motility, and Concomitant Phenomena, During Sleep." *Science* 118: 273–74.

Aspinall, Peter, Panagiotis Mavros, Richard Coyne, and Jenny Roe. 2013. "The Urban Brain: Analysing Outdoor Physical Activity with Mobile EEG." *British Journal of Sports Medicine* 49 (4): 1–6. doi:10.1136/bjsports-2012-091877.

Axon Sports. 2015. Accessed February 15. http://Axonsports.com.

Barry, Robert, et al. 2005. "Arousal and Activation in a Continuous Performance Task." *Journal of Psychophysiology* 19: 91–99.

Barry, Robert, et al. 2007. "EEG Differences between Eyes-Closed and Eyes-Open Resting Conditions." *Clinical Neurophysiology* 118: 2765–73.

"BBC Horizons Profiles Accelerated Learning: Golf." 2014. YouTube video, 5:43. Posted by AdvBrainMonitoring, December 23. www.youtube.com/watch?v=FaEqOcuxGow.

Beauchamp, Mark K., Richard Harvey, and Pierre Beauchamp. 2012. "An Integrative Biofeedback and Psychological Skills Training Program for Canada's Olympic Short-Track Speed Skating Team." *Journal of Clinical Sport Psychology* 6: 67–84.

Beaulieu, Anne. 2002. "Images Are Not the (Only) Truth: Brain Mapping, Visual Knowledge, and Iconoclasm." *Science, Technology and Human Values* 27 (1): 53–86.

———. 2004. "From Brainbank to Database: The Informational Turn in the Study of the Brain." *Studies in History and Philosophy of Science* 35 (2): 367–90.

Beck, Julie. 2014. "The Town That's Building Life around Sleep." *Atlantic*, February.

Benyus, Janine. 1997. *Biomimicry: Innovation Inspired by Nature*. New York: Harper Perennial.

Berger, Hans. (1929) 1969. "On the Electroencephalogram of Man." In *Hans Berger on the Electroencephalogram of Man: The Fourteen Original Reports on the Human Electroencephalogram*, translated and edited by Peter Gloor. New York: Elsevier.

Berka, Chris, Adrienne Behneman, Natalie Kintz, Robin Johnson, and Giby Raphael. 2010. "Accelerating Training Using Interactive Neuro-Educational Technologies: Applications to Archery, Golf and Rifle Marksmanship." *International Journal of Sport and Society* 1 (4): 87–104.

Berka, Chris, Gregory Chung, and Sam Nagashima. 2008. "Using Interactive Neuro-Educational Technology to Increase the Pace and Efficiency of Rifle Marksmanship Training." Presented at the Annual Meeting of the American Educational Research Association, New York, March.

Berson, Josh. 2014. *Computable Bodies: Instrumented Life and the Human Somatic Niche.* New York: Bloomsbury.

Bhanoo, Sindya N. 2014. "When Wearable Tech Saves Your Life, You Won't Take It Off." *Fast Company*, July 23. www.fastcompany.com/3033417/when-wearable-tech-saves-your-life-you-wont-take-it-off#4.

Biocybernaut Institute. 2015. Accessed May 25. www.biocybernaut.com/biocybernaut/.

———. 2016. "Biocybernaut Pricing." Accessed July 8. www.biocybernaut.com/pricing/.

"The Biofeedback & Neurofeedback at the Vancouver Olympics 2010 Secret." 2010. YouTube video, 5:58. Posted by Hal Myers, February 21. www.youtube.com/watch?v=sSXMGDpxYxE.

Blair, Joseph H. 1901. "Development of Voluntary Control." *Psychological Review* 8: 474–510.

Bonner, Geraldine. 1890. "In the Haworth." *Harper's Magazine* 80: 740–55.

Borck, Cornelius. 2001. "Electricity as a Medium of Psychic Life: Electrotechnological Adventures into Psychodiagnosis in Weimar Germany." *Science in Context* 14 (4): 565–90.

———. 2005. *Elektroenzephalographie* [The Emergence of the Electric Brain: A Cultural History of Electroencephalography]. Göttingen: Wallstein Verlag.

———. 2008. "Recording the Brain at Work: The Visible, the Readable, and the Invisible in Electroencephalography." *Journal of the History of Neurosciences* 17: 367–79.

———. 2016. "How We May Think: Imaging and Writing Technologies across the History of the Neurosciences." *Studies in History and Philosophy of Biological and Biomedical Sciences* 57: 112–20. http://dx.doi.org/10.1016/j.shpsc.2016.02.006.

Boxall, Andy. 2015. "These Headphones Won't Just Help You Sleep Better, They May Help Improve Your Memory." *Digital Trends*, June 21. www.digitaltrends.com/sports/kokoon-headphones-memory-improvements/.

Boyd Davis, Stephen. 2009. "Mapping the Unseen: Making Sense of the Subjective Image." In *Emotional Cartography: Technologies of the Self*, edited by Christian Nold, 38–52. www.emotionalcartography.net.

Boyd Davis, Stephen, Magnus Moar, Rachel Jacobs, Matt Watkins, and Mauricio Capra. 2007. "Mapping Inside Out." In *Pervasive Gaming Applications: A Reader for Pervasive Gaming Research, Volume 2*, edited by Carsten Magerkurth and Carsten Röcker, 199–226. Aachen: Shaker.

"Brainwaves for Business." 2015. *NHK World News*, April 3. www3.nhk.or.jp/nhkworld/english/news/techtrends/20150403.html.

Brazier, Mary A. B. 1984. "Pioneers in the Discovery of Evoked Potentials." *Electroencephalography and Clinical Neurophysiology* 59: 2–8.

Brenninkmeijer, Jonna. 2010. "Taking Care of One's Brain: How Manipulating the Brain Changes People's Selves." *History of the Human Sciences* 23 (1): 107–26.

———. 2013a. "Neurofeedback as a Dance of Agency." *BioSocieties* 8 (2): 144–63.
———. 2013b. "Brain Technologies of the Self: How Working on the Self by Working on the Brain Constitutes a New Way of Being Oneself." PhD diss., University of Groningen. http://irs.ub.rug.nl/ppn/35560826X.
———. 2016. *Neurotechnologies of the Self: Mind, Brain, and Subjectivity.* New York: Springer.
Brown, Chip. 2003. "The Stubborn Scientist Who Unraveled a Mystery of the Night." *Smithsonian* 34 (7): 92–100.
Browning, Norma Lee. 1956. "What Do Your Dreams Mean?" *Chicago Daily Tribune,* August 26.
Brunswick, David. 1924. "The Effects of Emotional Stimuli on the Gastro-Intestinal Tone: II. Results and Conclusions." *Journal of Comparative Psychology* 4: 225–87.
Burkus, David. 2013. "Are You Wasting Your 10,000 Hours?" *Forbes,* September 25. www .forbes.com/sites/davidburkus/2013/09/25/are-you-wasting-your-10000-hours/.
Canales, Jimena. 2001. "Exit the Frog, Enter the Human: Physiology and Experimental Psychology in Nineteenth-Century Astronomy." *British Journal for the History of Science* 34 (2): 173–97.
"Careers Could Ride on Brain Waves." 1972. *Nation's Business* 60 (6): 58–61.
Caton, Richard. 1875. "The Electric Currents of the Brain." *British Medical Journal* 2: 278.
CDRH (Center for Devices and Radiological Health). 2016. "General Wellness: A Policy for Low Risk Devices." Washington, DC: Food and Drug Administration, July 29. www .fda.gov/downloads/medicaldevices/deviceregulationandguidance/guidancedoc uments/ucm429674.pdf.
"Changes Colors with Your Moods." 1970. *TV Week* (Australia), n.d.
Choudhury, Suparna, and Jan Slaby, eds. 2011. *Critical Neuroscience: A Handbook of the Social and Cultural Contexts of Neuroscience.* New York: Wiley-Blackwell.
Clarke, Bruce, and Linda Henderson, eds. 2002. *From Energy to Information: Representation in Science and Technology, Art, and Literature.* Redwood City, CA: Stanford University Press.
"Coal & Allied Fighting Fatigue with the SmartCap." 2014. YouTube video, 3:57. Posted by NSW Mining, May 16. www.youtube.com/watch?v=P6V1Z_qTAII.
Conductar.com. 2015. "Moogfest 2014." Accessed August 15. www.conductar.com/#about.
Correal, Annie, and Andy Newman. 2014. "Head Trip." *New York Times,* June 10. http:// cityroom.blogs.nytimes.com/2014/06/10/new-york-today-head-trip.
Coverly, Merlin. 2006. *Psychogeography.* Ebbw Vale: CPD, Ltd.
Craciun, Mariana. 2013. "Embodied Expertise: The Science and Affect of Psychotherapy." PhD diss., University of Michigan. https://deepblue.lib.umich.edu/bitstream/handle /2027.42/102492/mcraciun_1.pdf.
Craig, Ashley, Yvonne Tran, Nirupama Wijesuriya, and Hung Nyuyen. 2012. "Regional Brain Wave Activity Changes Associated with Fatigue." *Psychophysiology* 49 (4): 574–82. doi:10.1111/j.1469-8986.2011.01329.x.
Craig, Melvin. 2014. "Bike Helmet Reads Your Mind as You Ride." MSNBC, June 21. www.msnbc.com/msnbc-live/watch/bike-helmet-reads-your-mind-as-you-ride -286040643714.
CRCMining. 2015a. "SmartCap Gains Traction with Miners." Accessed January 23. www .crcmining.com.au/news-and-media/newsroom/smartcap-gains-traction-with-miners/.
———. 2015b. "SmartCap: Operator Fatigue Management System." *CRCMining.* Accessed January 26. www.crcmining.com.au/breakthrough-solutions/smart-cap/.
Czeisler, Charles A. 1995. "The Effect of Light on the Human Circadian Pacemaker." *Ciba Foundation Symposium* 183: 254–302.

Daston, Lorraine, and Peter Galiston. 2004. *Objectivity.* Cambridge, MA: MIT Press.
Daston, Lorraine, and Elizabeth Lunbeck. 2011. "Introduction: Observation Observed." In *Histories of Scientific Observation,* edited by Lorraine Daston and Elizabeth Lunbeck. Chicago: University of Chicago Press, 2011.
Davy, John. 1967. "Sleep." *Daily Observer,* August 13.
Debord, Guy. (1955) 2015. "Introduction to a Critique of Urban Geography." Translated by Ken Knabb. In *The Situationist International Online.* Accessed August 15. www.cddc.vt .edu/sionline/presitu/geography.html.
———. (1956) 2006. "Theory of the Dérive." In *The Situationist International Anthology,* edited by Ken Knabb, 62–66. Berkeley: Bureau of Public Secrets.
Deleuze, Gilles. 1992. "Postscript on the Societies of Control." *October* 59: 3–7.
Demarest, Marc. 2016. "A Curious and Thoughtful Letter: Brain-Waves, the Metaphysical Society, 1869." *Chasing Down Emma* (blog), January. http://ehbritten.blogspot.com /2016/01/a-curious-and-thoughtful-letter-brain.html.
Dement, William C. 2004. "The Paradox of Sleep: The Early Years." *Archives Italiennes de Biologie* 142: 333–45.
Dement, William C., and Nathaniel Kleitman. 1957. "Cyclic Variations in EEG during Sleep and Their Relation to Eye Movements, Body Motility and Dreaming." *Electroencephalography and Clinincal Neurophysiology* 9: 673–90.
Derickson, Alan. 2013. *Dangerously Sleepy: Overworked Americans and the Cult of Manly Wakefulness.* Philadelphia: University of Pennsylvania Press.
Diamond, Edwin. 1962a. "The Science of Dreams—Fifth Article in a Series." *New York Herald Tribune,* April 8.
———. 1962b. "The Science of Dreams." *New York Herald Tribune,* April 15.
———. (1962) 1968. *The Science of Dreams.* New York: MacFadden-Bartell.
Dick, Philip K. 1997. "We Can Remember It for You Wholesale." In *The Philip K. Dick Reader.* Secaucus, NJ: Carol Publishing Group, 1997.
"Discovery Science's: Through the Wormhole with Morgan Freeman—Advanced Brain Monitoring Technology." 2013. YouTube video, 6:09. Posted by AdvBrainMonitoring, August 8. www.youtube.com/watch?v=udy_XzjE9Ls.
Dow Schüll, Natasha. 2016a. "Data for Life: Wearable Technology and the Design of Self-Care." *Biosocieties*: 1–17.
———. 2016b. "Sensor Technology and the Time-Series Self." *Continental Connections* 5 (1): 24–28.
Dror, Otniel E. 1999. "The Scientific Image of Emotion: Experience and Technologies of Inscription." *Configurations* 7: 355–401.
———. 2011. "Seeing the Blush: Feeling Emotions." In *Histories of Scientific Observation,* edited by Lorraine Daston and Elizabeth Lunbeck, 326–48. Chicago: University of Chicago Press.
Ducao, Arlene. 2014. *MindRider Maps Manhattan.* Brooklyn, NY: DuKorp.
DuKode Studio. 2015. "MindRider." Accessed August 15. http://dukodestudio.com/Mind Rider/.
DuKorp. 2015. "What Is MindRider?" Accessed August 15. http://mindriderhelmet.com.
Dumit, Joseph. 1999. "Objective Brains, Prejudicial Images." *Science in Context* 12 (1): 173–201.
———. 2003. "Is It Me or My Brain? Depression and Neuroscientific Facts." *Journal of Medical Humanities* 24 (1–2): 35–47.
———. 2004. *Picturing Personhood: Brainscans and Biomedical Identity.* Princeton: Princeton University Press.

Duncan, David. 2012. "A Little Device That's Trying to Read Your Thoughts." *New York Times*, April 2. www.nytimes.com/2012/04/03/science/ibrain-a-device-that-can-read -thoughts.html.

"8-Hour Sleep Myth Debunked? New Study Shows Hunter-Gatherer Tribes Get Less Shut-Eye." 2015. *RT*, October 16. www.rt.com/news/318893-study-preindustrial-society -sleep/.

Ekirch, A. Roger. 2001. "Sleep We Have Lost: Pre-industrial Slumber in the British Isles." *American Historical Review* 106 (2): 343–86.

———. 2005. *At Day's Close: Night in Times Past*. New York: Norton.

Ericsson, K. Anders, and Neil Charness. 1994. "Expert Performance: Its Structure and Acquisition." *American Psychologist* 49 (8): 725–47. doi:10.1037/0003-066X.49.8.725.

Ericsson, K. Anders, Ralf T. Krampe, and Clemens Tesch-Römer. 1993. "The Role of Deliberate Practice in the Acquisition of Expert Performance." *Psychological Review* 100 (3): 363–406. doi:10.1037/0033-295X.100.3.363.

Fermaglich, Joseph. 1971. "Determining Cerebral Death." *American Family Physician* 3: 85–87.

Fitzgerald, Des, Nikolas Rose, and Ilina Singh. 2016. "Living Well in the Neuropolis." *Sociological Review Monographs* 64 (1): 221–37. doi:10.1111/2059-7932.12022.

Forster, John. 2011. "Australia: CRC Mining and SmartCap." Updated March 16. www .mondaq.com/australia/x/125698/Trademark/CRCMining+and+SmartCap.

Fortunati, Leopoldina, James E. Katz, and Raimonda Riccini, eds. 2003. *Mediating the Human Body: Technology, Communication, and Fashion*. Mahwah, NJ: Lawrence Erlbaum.

Foucault, Michel. 1995. *Discipline and Punish: The Birth of the Prison*, translated by Alan Sheridan. New York: Vintage Books.

Foulkes David. 1966. *The Psychology of Sleep*. New York: Scribner.

———. 1996. "Sleep and Dreams, Dream Research 1953–1993." *Sleep* 19 (8): 609–24.

Franke, Jeane. 1958. "Tribune Girl Reporter Holds Down Dream Job." *Chicago Tribune*, August 24.

Franklin, Sarah, Celia Lury, and Jackie Stacey. 2004. *Global Nature, Global Culture*. Thousand Oaks, CA: SAGE Publishing.

Freake, Douglas. 1995. "The Semiotics of Wristwatches." *Time and Society* 4 (1): 67–90.

Freedman, John Craig. 2015. "EEG AR: Things We Have Lost." *John Craig Freedman* (blog). Accessed August 15. https://johncraigfreeman.wordpress.com/eeg-ar-things-we -have-lost/.

Frick, Laurie. 2014. "Self-Surveillance: Should You Worry or Simply Embrace Your Personal Data?" *Science and Society*, February 7, 218–22. doi:10.1002/embr.201438460.

Gamble, Jessa. 2011. *Siesta and the Midnight Sun: How Our Bodies Experience Time*. New York: Viking.

Gardner, Paula, and Britt Wray. 2013. "From Lab to Living Room: Transhumanist Imaginaries of Consumer Brain Wave Monitors." *Ada: A Journal of Gender, New Media, and Technology* 3. doi:10.7264/N3GQ6VP4.

Geddes, Leslie. 2000. "Historical Perspectives 5: Electroencephalography." In *Biomedical Engineering Handbook*, 2nd ed., edited by Joseph Bronzino. Berlin: Springer-Verlag.

Giaccardi, Elisa, and Daniela Fogli. 2008. "Affective Geographies: Toward a Richer Cartographic Semantics for the Geospatial Web." *AVI'08*, May 28–30, 173–80.

Gibbons, Roy. 1964. "Brain Signals Death of Man, Doctor Says." *Chicago Tribune*, June 24.

Gibbs, Frederic, and Erna Gibbs. 1941. *Atlas of Electroencephalography*. New York: Lew A. Cummings.

Ginzberg, Raphael. 1949. "Three Years with Hans Berger: A Contribution to His Biography." *Journal of the History of Medicine and Allied Sciences* 4 (4): 361–71.

Gloor, Peter. 1969. *Hans Berger: On the Electroencephalogram of Man*. Preface by Herbert H. Jasper. New York: Elsevier.

———. 1994. "Berger Lecture. Is Berger's Dream Coming True?" *Electroencephalography and Clinical Neurophysiology* 90: 253–66.

Goodenough, Donald, Arthur Shapiro, Melvin Holden, and Leonard Steinschhiber. 1959. "A Comparison of 'Dreamers' and 'Non Dreamers': Eye Movements, Electroencephalograms, and the Recall of Dreams." *Journal of Abnormal and Social Psychology* 59: 295–302.

Gottesmann, Claude. 2001. "The Golden Age of Rapid Eye Movement Sleep Discoveries: Lucretius—1964." *Progress in Neurobiology* 65: 211–87.

———. 2005. *The Golden Age of Rapid Eye Movement Sleep Discoveries: 1965–1966*. New York: Nova Science Publishers.

Grau, Charles, et al. 2014. "Conscious Brain-to-Brain Communication in Humans Using Non-Invasive Technologies." *PLOS One*, August 19. http://journals.plos.org/plosone /article?id=10.1371/journal.pone.0105225#pone-0105225-g001.

Greenberg, Joshua M. 2004. "Creating the 'Pillars': Multiple Meanings of a Hubble Image." *Public Understanding of Science* 13: 83–95.

GuardVant. 2015. "OPGUARD: How It Works." Accessed January 26. www.guardvant .com/solution/opguard/opguard-how-it-works/.

Guljajeva, Vavarna, M. Canet, and S. Mealla. 2012. "Neuro Knitting." *Knitic*. Accessed January 28. www.varvarag.info/neuroknitting.

Gustaitis, Rasa. 1971. "The Alpha Gambit." *Los Angeles Times*, August 8.

Hall, Rachel. 2007. "Of Ziplock Bags and Black Holes: The Aesthetics of Transparency in the War in Terror." *Communication Review* 10 (4): 319–46. doi:10.1080/10714420701715381.

———. 2015a. *The Transparent Traveler: The Performance and Culture of Airport Security*. Durham, NC: Duke University Press.

———. 2015b. "Terror and the Female Grotesque: Introducing Full-Body Scanners to U.S. Airports." In *Feminist Surveillance Studies*, edited by Rachel E. Dubrofsky and Shoshana Magnet, 127–49. Durham, NC: Duke University Press.

Hamilton, J. R. 1912. "Trained Muscles vs. Trained Brains." *San Francisco Chronicle*, August 25.

Harris, Jonathan, and Sepandar Kamvar. 2006. "Mission." We Feel Fine, May. http:// wefeelfine.org/mission.

Hegarty, Stephanie. 2012. "The Myth of the Eight-Hour Sleep." *BBC News Magazine*, February 22. www.bbc.com/news/magazine-16964783.

Hubel, D. H. 1960. "Electrocotiograms in Cats during Natural Sleep." *Archives italiennes de biologie* 98: 171–81.

"The Hypothesis of Brain-Waves." 1869. *Spectator*, January 30. http://archive.spectator.co .uk/article/30th-january-1869/10/the-hypothesis-of-brain-waves.

Intaglia, Christopher. 2015. "'Brain Prints' Could Be Future Security ID." *Scientific American*, June 5. www.scientificamerican.com/podcast/episode/brainprints-could-be-future -security-id/.

"Italy's Weapon Is All in Their Heads." 2006. *Montreal Gazette*, July 8. www.canada.com /montrealgazette/news/story.html?id=f0289b54-acc1-4710-899d-e6227b80e88a.

Jasper, Herbert H., and Howard L. Andrews. 1936. "Human Brain Rhythms: I. Recording Techniques and Preliminary Results." *Journal of General Psychology* 14 (1): 98–126. doi:10.1080/00221309.1936.9713141.

Jennett, Bryan. 1975. "The Doctor's Dilemma." *Journal of Medical Ethics* 1: 63–66.

Jouvet, Michel. 1959. "Diagnostic électro-sous-cortico-graphique de la mort du système nerveux central au cours de certains comas." *Electroencephalography and Clinical Neurophysiology* 11 (4): 11805–8.

Joyce, Kelly. 2008. *Magnetic Appeal: MRI and the Myth of Transparency*. Ithaca: Cornell University Press.

Joyce, Kelly, and Meika Loe. 2010. *Technogenarians: Studying Health and Illness through an Aging, Science, and Technology Lens*. New York: Wiley.

Kamiya, Joe. 1968. "Conscious Control of Brain Waves." *Psychology Today* 1 (11): 56–60.

———. 1971. "Operant Control of the EEG Alpha Rhythm and Some of Its Reported Effects on Consciousness." In *Biofeedback and Self-Control: An Aldine Reader on the Regulation of Bodily Processes and Consciousness*, edited by C. T. Tart, 507–17. New York: Wiley.

Kansara, Vikram Alexei. 2014. "Amanda Parkes on Why Wearable Tech Is about More Than Gadgets." *Business of Fashion*, November 30. www.businessoffashion.com/2014/11/amanda-parkes-wearable-tech-gadgets.html.

Kaplan, Robert. 2011. "The Mind Reader: The Forgotten Life of Hans Berger, Discoverer of the EEG." *Australasian Psychiatry* 19 (2). doi:10.3109/10398562.2011.561495.

Kaufman, Josh. 2013. *The First 20 Hours: How to Learn Anything . . . Fast!* New York: Portfolio.

Khushaba, Rami, Chelsea Wise, Sarath Kodagoda, Jordan Louviere, Barbara E. Kahn, and Claudia Townsend. 2013. "Consumer Neuroscience: Assessing the Brain Response to Marketing Stimuli Using Electroencephalogram (EEG) and Eye Tracking." *Expert Systems with Applications* 40: 3803–12.

Kirkland, Kyle. 2013. *Mind Plague*. N.p.: Neuroniverse, 2013.

Knowles, James Thomas. 1869. "Brain-waves—A Theory." *Spectator*, January 20, 11. http://archive.spectator.co.uk/article/30th-january-1869/11/brain-wavesa-theory.

Kokoon. 2017. "Homepage." Accessed April 1. http://kokoon.io.

Koslofsky, Craig. 2011. *Evening's Empire: A History of the Night in Early Modern Europe*. Cambridge University Press.

Kostyukovsky, Nina. 2015. "Regulating Wearable Devices in the Healthcare Sector." *American Bar Association Health Law Section* 11 (9). www.americanbar.org/publications/aba_health_esource/2014-2015/may/devices.html.

Krassenstein, Eddie. 2014. "NEUROTiQ—3D Printed Headdress Monitors and Exhibits the States of the Wearer's Brain." *3Dprint*, September 14. https://3dprint.com/13409/NEUROTiQ-3d-printed-headdress/.

Kremin, Irwin. 1962. "And Perchance to Misinterpret Our Dreams: There's Another Rub." *New York Times*, May 6.

Kunz, Marine. 2014. "Brain Sensor Cap Illuminates Colors of Thought." *PSFK*, October 2. www.psfk.com/2014/10/NEUROTiQ-brain-sensor-cap-lights-up-fashion-week.html.

Kusenbach, Margarethe. 2003. "Street Phenomenology: The Go-Along as Ethnographic Research Tool." *Ethnography*, 4 (3): 455–85.

Laird, Donald A. 1937. *How to Sleep Better*. New York: Funk & Wagnalls.

Lakhtakia, Akhlesh, and Raul Jose Martin-Palma, eds. 2013. *Engineering Biomimicry*. Atlanta: Elsevier.

Lance, Brent, Scott E. Kerick, Anthony J. Ries, Kelvin S. Oie, and Keleb McDowell. 2012. "Brain-Computer Interface Technologies in the Coming Decades." *Proceedings of the IEEE* 100. doi:10.1109/JPROC.2012.2184830.

Latour, Bruno. 1987. *Science in Action: How to Follow Scientists and Engineers through Society.* Cambridge, MA: Harvard University Press.

Laureys, S., M. Boly, G. Moonen, and P. Maquet. 2009. "Coma." *Encyclopedia of Neuroscience* 2: 1133–42.

Lecher, Colin. 2015. "The FDA Doesn't Want to Regulate Wearables, and Device Makers Want to Keep It That Way." *TheVerge*, June 24. www.theverge.com/2015/6/24 /8836049/fda-regulation-health-trackers-wearables-fitbit.

Levine, Barry. 2014. "Tune Your Meditation with a Metal Flower That Tracks Your Brainwaves—and Those of Your Friends." *VBNews*, June 25. http://venturebeat.com /2014/06/25/tune-your-meditation-with-a-metal-flower-that-tracks-your-brainwaves -and-those-of-your-friends/.

Libet, Benjamin, C. A. Gleason, E. W. Wright, and D. K. Pearl. 1983. "Time of Consciousness Intention to Act in Relation to Onset of Cerebral Activity (Readiness-Potential): The Unconscious Initiation of Freely Voluntary Act." *Brain* 106 (3): 623.

Linden, David. 2014. *Brain Control: Developments in Theory and Implications for Society.* New York: Palgrave.

Littlefield, Melissa M. 2009. "Constructing the Organ of Deceit: The Rhetoric of fMRI and Brain Fingerprinting in Post-9/11 America." *Science, Technology & Human Values* 34: 365–92.

———. 2010. "Matter for Thought: The Psychon in Neurology, Psychology and American Culture, 1927–1943" *Neurology and Modernity.* Eds. Andrew Shail and Laura Salisbury. Houndsmills: Palgrave. 267–86.

———. 2011. *The Lying Brain: Lie Detection in Science and Science Fiction.* Ann Arbor: University of Michigan Press, 2011.

Littlefield, Melissa M., and Jenell Johnson. 2012. *The Neuroscientific Turn: Transdisciplinarity in the Age of the Brain.* Ann Arbor: University of Michigan Press.

Loomis, Alfred, Newton Harvey, and Garret Hobart. 1937. "Cerebral States during Sleep as Studied by Human Brain Potentials." *Journal of Experimental Psychology* 21: 127–44.

Lorimer, Jamie. 2010. "Moving Image Methodologies for More Than Human Geographies." *Cultural Geographies* 17 (2): 237–58.

"The Lotus: Bloom a Flower with Your Mind." 2014. Kickstarter project. Posted June 22 by Mindfulness, Inc. www.kickstarter.com/projects/mindfulness/the-lotus?ref=video.

Luce, Gay, and Eric Pepper. 1971. "Mind over Body, Mind over Mind." *New York Times Sunday Magazine*, September 12.

Luce, Gay Gaer, and Julius Segal. 1966a. *Sleep.* New York: Coward-McCann.

———. 1966b. "Sleep—From Alpha to Delta." *New York Times*, April 17.

Luckhurst, Roger. 2002. *The Invention of Telepathy.* Oxford: Oxford University Press.

Lupton, Deborah. 2016. *The Quantified Self.* Cambridge: Polity Press.

Machado, Calixto et al. 2007. "The Declaration of Sydney on Human Death." *Journal of Medical Ethics* 33: 699–703. doi:10.1136/jme.2007.020685.

MacKenzie, Norman. 1966. "That Dark Third of Life: Review." *New York Times*, June 12.

Maksel, Rebecca. 2008. "When Did the Term "Jet Lag" Come into Use?" *Air and Space Magazine*, June 17.

Malin, Brenton J. 2014. *Feeling Mediated: A History of Media Technology and Emotion in America.* New York: NYU Press.

Mann, Jeff. 2015. "Sleep Junkies Interview with David Cloud." *Sleep Junkies*, October 20. http://sleepjunkies.com/blog/national-sleep-foundation-interview-sleep-tech/.

———. 2016. "Sleep Junkies Interview with Tim Antos." *Sleep Junkies*, June 1. http://sleep junkies.com/blog/kokoon-eeg-headphones/.

Martin, Emily. 2010. "Self-Making and the Brain." *Subjectivity* 3 (4): 366–81.
Maslen, Hannah, Thomas Douglas, Roi Cohen Kadosh, Neil Levy, and Julian Savulescu. 2014. "The Regulation of Cognitive Enhancement Devices: Extending the Medical Model." *Journal of Law and the Biosciences* 1 (1): 68–93.
———. 2015. "The Regulation of Cognitive Enhancement Devices: Refining Maslen et al.'s Model." *Journal of Law and the Biosciences* 2 (3): 754–67. doi:10.1093/jlb/lsvo29.
Mathis, Johannes. 1995. "Abstract: The History of Sleep Research in the 20th Century" [article in German]. www.ncbi.nlm.nih.gov/pubmed/8539501.
Mavros, Panagiotis. 2011. "Emotional Urbanism." MS thesis, Edinburgh College of Art, University of Edinburgh.
McDowell, Linda. 1999. *Gender, Identity and Place: Understanding Feminist Geography.* Minneapolis: University of Minnesota Press.
"Measure Driver's Brain Waves." 1957. *Chicago Daily Tribune*, September 8.
Merriman, Peter. 2014. "Rethinking Mobile Methods." *Mobilities* 9 (2). doi:10.1080/17450 101.2013.784540.
Mihajlović, Vojkan, Bernard Grundlehner, Rudd Vullers, and Julien Penders. 2015. "Wearable, Wireless EEG Solutions in Daily Life Applications: What Are We Missing?" *IEEE Journal of Biomedical and Health Informatics* 19 (1): 6–21.
Millett, David. 2001. "Hans Berger: From Psychic Energy to EEG." *Perspectives in Biology and Medicine* 44 (4): 522–42.
Millington, Brad. 2011. "Use It or Lose It: Ageing and the Politics of Brain Training." *Leisure Studies* 31 (4): 429–46.
———. 2014. "Amusing Ourselves to Life: Fitness Consumerism and the Birth of Bio-Games." *Journal of Sport and Social Issues* 38 (6): 491–508.
"MindRider and the Maker's Brain." 2014. YouTube video, 21:15. Posted by Make:, September 22. www.youtube.com/watch?v=vr6Ks8Vrm3M.
Mindroom Peak Sports Performance. 2015a. "About Mindroom." Accessed May 5. www .mindroompsp.com/en-about.html.
———. 2015b. "Mindroom Home." Accessed March 19. www.mindroompsp.com/index .html.
———. 2015c. "Mindroom Summary." Accessed May 14. www.mindroompsp.com/Mi ndRoom-International.pdf.
———. 2015d. "Sport Science Frequently Asked Questions." Accessed February 15. http://mindroompsp.com/en-faq.html.
———. 2015e. "What Is Neurofeedback and Others." Accessed May 5. www.mindroompsp .com/en-faq.html#5.
Miranda, Robbin A., et al. 2015. "DARPA-Funded Efforts in the Development of Novel Brain–Computer Interface Technologies." *Journal of Neuroscience Methods* 244: 52–67. doi:10.1016/j.jneumeth.2014.07.019.
Mollaret, P., and M. Goulon. 1959. "Le coma dépassé mémoire préliminaire." *Revue neurologique* (Paris) 101 (1): 3–15.
"'Mood Sweater' | Sensoree Therapeutic Bio.Media." YouTube video, 2:20. Posted by Beyond-Print, May 1. www.youtube.com/watch?v=ow8cFhxvCaA.
Morgan, Alex. 2015. "People Who Are Revolutionizing Data in Sport." *FootBaller Everyday*, April 29. http://footballeveryday.co.uk/2015/04/29/the-people-who-are-revolu tionizing-data-in-sport/.
Morrison, Adrian R. 2013. "Coming to Grips with a 'New' State of Consciousness: The Study of Rapid-Eye-Movement Sleep in the 1960s." *Journal of the History of the Neurosciences* 22 (4): 392–407.

Munsterberg, Hugo. 1908. *On the Witness Stand: Essays on Psychology and Crime.* New York: Doubleday, Page.

Naïm, Omar (director). 2004. *Final Cut.* DVD, 95 min. Lions Gate Entertainment.

Necomimi. 2015. Accessed February 14. www.necomimi.com/.

Neff, Gina, and Dawn Nafus. 2016. *Self Tracking.* Cambridge, MA: MIT Press.

Nelson, Harry. 1965. "Oldsters Sleep Less, Brain Waves Hint." *Los Angeles Times,* December 9.

———. 1969. "'Flat' Brain Waves—the Sign of Death?" *Los Angeles Times,* September 15.

Nelson, Lise, and Joni Seager. 2004. *A Companion to Feminist Geography.* New York: Wiley.

Neuroon. 2016a. "Neuroon Home." Accessed March 15. https://neuroon.com/.

———. 2016b. "Neuroon Help." Accessed March 15. https://neuroon.com/help/.

NeuroSky. 2015a. "EEG Biosensors." Accessed February 16. http://neurosky.com/biosensors/eeg-sensor/.

———. 2015b. "The Lotus." Accessed May 20. http://store.neurosky.com/products/the-lotus.

———. 2016. "PuzzleBox Bloom." Accessed July 8. http://store.neurosky.com/products/the-lotus.

"NeuroTracker LowDown." 2011. YouTube video, 4:26. Posted by NeuroTracker1, June 2. www.youtube.com/watch?v=H9EgV-FaM3M.

"NeuroTracker Sport Science Innovations Croatia." 2015. YouTube video, 2:03. Posted by Sport Science Innovations Croatia, March 15. www.youtube.com/watch?v=Q6GWARIby_c.

NeuroVigil. 2016. "FAQ." Accessed June 1. http://neurovigil.com/index.php/technology/faq.

Neurowear. 2015a. "FAQ." Accessed January 23. www.neurowear.com/faq/.

———. 2015b. "Projects / Neurocam." Accessed June 15. http://neurowear.com/projects_detail/neurocam.html.

———. 2016. "About Neuro Tagging Map." Accessed August 15. http://neurowear.com/projects_detail/neuro_tagging_map.html.

Nold, Christian, ed. 2009. *Emotional Cartography: Technologies of the Self.* www.emotional-cartography.net.

"Notes of the Month." 1935. *Discovery: A Monthly Popular Journal of Knowledge* 16 (181): 1.

Nowland, David. 1970. "Time-Zone's Effect on the Jet Set." *Irish Times,* May 2.

Otis, Laura. 2001. *Networking: Communicating with Bodies and Machines in the Nineteenth Century.* Ann Arbor: University of Michigan Press.

Ou, Po-Hsiang, and Skye Liu. 2016. "Wearables in Taiwan: A New Regulatory Category?" *Eiger Law.* www.eiger.law/en/publications/doc_details/332-wearables-in-taiwan-a-new-regulatory-category.

Pavón-Cuéllar, David. 2014. "Extimacy." In *Encyclopedia of Critical Psychology,* edited by Thomas Teo, 661–64. New York: Springer.

Pearson, Jordan. 2014. "Scientists Found a Way to Email Brainwaves." *Motherboard,* August 20. http://motherboard.vice.com/read/scientists-found-a-way-to-email-brainwaves.

Pickersgill, M., S. Cunningham-Burley, and P. Martin. 2011. "Constituting Neurologic Subjects: Neuroscience, Subjectivity and the Mundane Significance of the Brain." *Subjectivity* 4: 346–65.

Piwek L., D. A. Ellis, S. Andrews, and A. Joinson. 2016. "The Rise of Consumer Health Wearables: Promises and Barriers." *PLOS Medicine* 13 (2): e1001953. doi:10.1371/journal.pmed.1001953.

Pribram, Karl H., and Diane McGuinness. 1975. "Arousal, Activation, and Effort in the Control of Attention." *Psychological Review* 82 (2): 116–49.

———. 1992. "Attention and Para-Attentional Processing. Event-Related Brain Potentials as Tests of a Model." *Annals of the New York Academy of Sciences* 658: 65–92.

"Psychologists Get Some New Ideas of Stuff That Dreams Are Made Of." 1960. *Washington Post*, June 8.

"Putting a Cap on Fatigue in Mining." 2013. YouTube video, 3:58. Posted by NSW Mining, May 19. YouTube. www.youtube.com/watch?v=K_lQMBNhxYU.

"Q&A: Kokoon—Sleep Headphones That Read Your Brainwaves." 2015. *Sleep Junkies Blog*, June 1. http://sleepjunkies.com/blog/kokoon-eeg-headphones/.

Qing Guo, Jingxian Wu, and Baohua Li. 2014. "EEG-Based Golf Putt Outcome Prediction Using Support Vector Machine." *Computational Intelligence in Brain Computer Interfaces.* 2014 IEEE Symposium, Orlando, FL, December 9–12. doi:10.1109/CIBCI .2014.7007790.

Rechtschaffen, Allan. 1998. "Current Perspectives on the Function of Sleep." *Perspectives in Biology and Medicine* 41 (3): 359–90.

Rechtschaffen, A., B. M. Bergmann, C. A. Everson, C. A. Kushida, and M. A. Gilliland. 1989. "Sleep Deprivation in the Rat: X. Integration and Discussion of the Findings." *Sleep* 12: 68–87.

Ridge, Pett. 1917. *The Amazing Years*. London: Hodder & Stoughton.

Robbins, Jim. 2000. *A Symphony in the Brain: The Evolution of the New Brain Wave Biofeedback*. New York: Atlantic Monthly Press.

Robinson, Leonard Wallace. 1959. "What We Dream—And Why." *New York Times Sunday Magazine*, February 15.

Rose, Gillian. 1993. *Feminism and Geography: The Limits of Geographical Knowledge*. Minneapolis: University of Minnesota Press.

Rose, Nikolas. 2003. "Neurochemical Selves." *Society* 41 (1): 46–59.

Rose, Nikolas, and Joelle Abi-Rached. 2013. *Neuro: The New Brain Sciences and the Management of the Mind*. Princeton: Princeton University Press.

Saunders, Barry F. 2008. *CT Suite: The Work of Diagnosis in the Age of Noninvasive Cutting*. Durham, NC: Duke University Press.

Schiebinger, Londa. 2004. "Why Mammals Are Called Mammals." *Nature's Body: Gender and the Making of Modern Science*. New Brunswick, NJ: Rutgers University Press.

Schiff, Nicholas, and Fred Plum. 2000. "The Role of Arousal and 'Gating' Systems in the Neurology of Impaired Consciousness." *Journal of Clinical Neurology* 17 (5): 438–52.

Schirmann, Felix. 2014. "'The Wondrous Eyes of a New Technology': A History of the Early Electroencephalography (EEG) of Psychopathy, Delinquency, and Immorality." *Frontiers in Human Neuroscience*, April 17. doi:10.3389/fnhum.2014.00232.

Sensoree. 2014. "NEUROTiQ: Artifact." Accessed December 14. http://sensoree.com /artifacts/NEUROTiQ/.

———. 2015. "About Sensoree." Accessed February 13. http://sensoree.com/about/.

———. 2016. "FURVER fo.corset: Artifact." Accessed July 10. http://sensoree.com /artifacts/furver/.

Sharpley, Christopher Francis, and Vicki Bitsika. 2010. "The Diverse Neruogeography of Emotional Experience: Form Follows Function." *Behavioural Brain Research* 215: 1–6. doi:10.1016/j.bbr.2010.06.031.

Shepard, John W., Daniel Buysse, Andrew Chesson Jr., William C. Dement, et al. 2005. "History of the Development of Sleep Medicine in the United States." *Journal of Clinical Sleep Medicine* 1 (1): 61–82.

Sherlin, Leslie, Noel Larson, and Rebecca Sherlin. 2013. "Developing a Performance Brain Training Approach for Baseball: A Process Analysis with Descriptive Data." *Applied Psychophysiology and Biofeedback* 38 (1): 29–44. doi:10.1007/s10484-012-9205-2.

SmartCap. 2015a. "FAQ SmartCap." Accessed February 14. www.smartcaptech.com/faq/.

———. 2015b. "Overview." Accessed February 14. www.smartcaptech.com/our-product/.

Smith, William. 1972. "Can Man Control His Mind?" *New York Times*, April 16.

Sourkes, T. L. 2006. "On the Energy Cost of Mental Effort." *Journal of the History of the Neurosciences* 15 (1): 31–47.

Spinney, Justin. 2015. "Close Encounters? Mobile Methods, (post)Phenomenology and Affect." *Cultural Geographies* 22 (2): 231–46.

Squier, Susan. 2004. *Liminal Lives: Imagining the Human at the Frontiers of Biomedicine.* Durham, NC: Duke University Press.

"Startup of the Year 2015–2016 Competition Entry: Kokoon Technology." 2016. *Guardian*, February 10. www.theguardian.com/small-business-network/2016/feb/10/startup -of-the-year-competiton-entry-kokoon-technology.

Stafford, Tom. 2009. "Hacking Our Tools for Thought." In *Emotional Cartographies: Technologies of the Self,* edited by Christian Nold, 88–95. www.emotionalcartography .net.

Stearns, Peter. 1994. *American Cool: Constructing a Twentieth-Century Emotional Style.* New York: NYU Press.

Stinson, Bruce, and David Arthur. 2013. "A Novel EEG for Alpha Brain State Training, Neurofeedback and Behavior Change." *Complementary Therapies in Clinical Practice* 19: 114–18.

Sullivan, Meg. 2015. "Our Ancestors Probably Didn't Get 8 Hours of Sleep Either." *UCLA Newsroom*, October 15. http://newsroom.ucla.edu/releases/our-ancestors-probably -didnt-get-8-hours-a-night-either.

Sung, Dan. 2015. "Wearable Tech and Regulation: Should Fitness Trackers Face the FDA?" *Wareable*, September 24. www.wareable.com/health-and-wellbeing/wearable -tech-and-regulation-5678.

Sutton, Horace. 1966. "The Jet Lag Leaves Travelers up in the Air." *Los Angeles Times*, Febrary 20.

Tan, Desney, and Anton Nijholt. 2010. *Brain-Computer Interfaces: Applying Our Minds to Human-Computer Interaction.* New York: Springer.

Thornton, Davi Johnson. 2011. *Brain Culture: Neuroscience and Popular Media.* New York: Routledge.

Thought Technology Ltd. 2017. "About Us." Accessed April 1. http://thoughttechnology. com/index.php/about-us.

Thrift, Nigel. 2004. "'Intensities of Feeling': Towards a Spatial Politics of Affect." *Geografisker Annaler* 86B (1): 57–78.

Ubell, Earl. 1962. "The Mind's Late Late Show." *New York Herald Tribune*, April 1.

Umapathy, Kiran. 2015. "Noise-Cancelling Headphones Were Designed to Facilitate Sleep." *PSFK*, May 19. www.psfk.com/2015/05/noise-cancelling-headphones-for-sleep -noise-cancelling.html.

Underman, Kelly. 2015. "Playing Doctor: Simulation in Medical School as Affective Practice." *Social Science and Medicine* 136–37: 180–88.

"Using Wearable Technology to Read Your Moods." 2013. YouTube video, 20:35. Posted by Microsoft Researchers, November 22. www.youtube.com/watch?v=b-jpDy8loEg.

VaezMousavi, S. M., Robert J. Barry, Jacqueline A. Rushby, and Adam R. Clarke. 2007. "Evidence for Differentiation of Arousal and Activation in Normal Adults." *Acta Neurobiologiae Experimentalis* 67 (2): 179–86.

Van Dellen, Theodore. 1956. "How to Keep Well: Sleep Waves?" *Washington Post*, August 9.

———. 1960. "How to Keep Well: Dream Boys." *Chicago Daily Tribune*, August 14.

———. 1969. "How to Keep Well: Definition of Death." *Chicago Tribune*, April 21.

Van Der Drift, Marcel. 2009. "A Future Love Story." In *Emotional Cartography: Technologies of the Self*, edited by Christian Nold, 27–32. www.emotionalcartography.net.

Vennare, Joe. "What to Become a Better Athlete? Train Your Brain." *DailyBurn*. Accessed 5/5/15. http://dailyburn.com/life/tech/brain-training-sports/.

Vidal, Fernando. 2009. "Brainhood, Anthropological Figure of Modernity." *History of the Human Sciences* 22 (1): 5–36.

Viney, Michael. 1966. "No Perchance about It." *Irish Times*, January 29.

Waldby, Catherine. 2000. *The Visible Human Project: Informatic Bodies and Posthuman Medicine*. New York: Routledge.

Walker, Tim. 2013. "SmartCap Heads into the Blue." *GizMag*, September 8. www.gizmag.com/smartcap-marine-trials/28873/.

"Wearables+Tech: Kristin Neidlinger." 2014. YouTube video, 5:42. Posted by Institute for the Future, March 18. www.youtube.com/watch?v=PFdEbN41Tks.

Wehr, Thomas. 1996. "A 'Clock for All Seasons' in the Human Brain." *Progress in Brain Research* 111: 321–42.

Wertheimer, P., M. Jouvet, and J. Descots. 1959. "Diagnosis of Death of the Nervous System in Comas with Respiratory Arrest Treated by Artificial Respiration." *La presse médicale* 67 (3): 87–88.

Wexler, Anna. 2015. "A Pragmatic Analysis of the Regulation of Consumer Transcranial Direct Current Stimulation (tDCS) Devices in the United States." *Journal of Law and Biosciences* 2 (3): 669–96. doi:10.1093/jlb/lsvo39.

"What Is Neurofeedback?—EEG INFO Videos." 2007. YouTube video, 8:33. Posted by Kurt Othmer, April 8. www.youtube.com/watch?v=t6XeCwFQrCA.

"What's Next—A Window on the Brain: Chris Berka at TEDxSanDiego 2013." 2014. YouTube video, 8:51. Posted by TEDx Talks, February 4. www.youtube.com/watch?v=rBt7LmrIkxg.

Williams, Simon J., Paul Higgs, and Stephen Kats. 2012. "Neuroculture, Active Ageing and the 'Older Brain': Problems, Promises and Prospects." *Sociology of Health & Illness* 34 (1): 64–78.

Wilson, Elizabeth. 2011. "Neurological Entanglements: The Case of Paediatric Depressions, SSRIs and Suicidal Ideation." *Subjectivity* 4: 277–97.

Wilson, Vietta, Erik Peper, and Donald Moss. 2011. "'The Mind Room' in Italian Soccer Training: Use of Biofeedback and Neurofeedback for Optimum Performance." *Biofeedback* 34 (3): 79–81.

Wolpaw, Jonathan R., Niels Birbaumer, Dennis J. McFarland, Gert Pfurtscheller, and Theresa M. Vaughan. 2002. "Brain-Computer Interfaces for Communication and Control." *Clinical Neurophysiology* 113: 767–91.

Wolpaw, Jonathan R., and Elizabeth Winter Wolpaw. 2012. "Brain-Computer Interfaces: Something New under the Sun." In *Brain-Computer Interfaces: Principles and Practice*, edited by Jonathan R. Wolpaw and Elizabeth Winter Wolpaw, 3–12. Oxford: Oxford University Press.

Yetish, Gandhi, et al. 2015. "Natural Sleep and Its Seasonal Variations in Three Pre-Industrial Societies." *Current Biology* 25: 2862–68.

Zhang, Biao, Jianjun Wang, and Thomas Fuhlbrigge. 2010. "Review of the Commercial Brain-Computer Interface Technology from Perspective of Industrial Robotics." *Proceedings of the 2010 IEEE International Conference on Automation and Logistics, Hong Kong and Macau, August 16–20*: 379–84.

Index

Rechtschaffen, Allan, 73, 84, 95, 136n5
regulation: bodily, 47, 53, 62, 72; medical
devices and, 8–9, 132n21
REM. *See* rapid eye movement
responsibility, 16, 49–50, 52–53, 58, 128,
132n24
risk: discourses of, 3, 17, 21, 57–58, 128,
132n21; government regulation and, 8–9
Robbins, Jim, 52–56, 61, 134n5, 135n8
Rose, Nikolas, 6, 15, 16–17, 49–51, 53,
57, 121

Science and Technology Studies, 40, 107,
132n21
science fiction, 4, 16, 120, 131n16
self: conceptions of, 14–15, 19, 49–51, 115;
knowledge of, 34, 47, 107, 113; machine
interface for, 18, 32; mastery of, 65–66,
68–69
self-tracking, 3–4, 18–19, 24–25, 97, 132n24
senses, 7, 36, 54
Sensoree, 24, 30–33, 36, 133n2
sensory processing disorder, 28, 43
shift work, 94–95. *See also* sleep
Siegel, Jerome, 72–74
signal quality, 7, 42
Situationists, 21, 99–102
Slaby, Jan, 129n1
sleep, 2–3, 95, 136n3; and conscious or
unconscious state, 13, 71–72, 76–79,
85–86; delta waves and, 41, 43; measurement of, 76, 89, 136n5; naturalness of,
71–76; patterns of, 73–75, 135n1;
psychoanalysis and, 83–85; quality and
optimization of, 86–88, 91–95. *See also*
chronotype; dreaming; Kokoon; Neuroon;
rapid eye movement
sleep disorders, 89–90
sleep learning, 87
SmartCap, 2, 29, 37–39, 41, 43–46, 125,
133nn7–8
smartphone: and disrupting natural
rhythms, 71; interface with, for EEG, 8–9,
69, 96, 106, 134n13; and Neurocam, 98,
113–116; and sleep apps, 92, 94
speculative fiction, 20
sport psychology, 58–59, 61
Stafford, Tom, 121

states of mind: correlations between
physiology and, 9–10, 42, 47, 83–84, 95,
107; creation of, 3–4, 17, 124–125, 127;
desirable, 24, 27, 43–46
Stearns, Peter, 9, 46
Sternman, Barry, 61, 135n5
stress, 9, 25, 88, 107, 110, 121; arousal as, 32,
122; fatigue and, 47; Inflatable Corset and,
30; recalibration of, 27; training to reduce,
58–59
STS. *See* Science and Technology Studies
surveillance, 20, 24, 27, 29, 37–39

technological mediation, 14, 18, 23, 32, 91–92
telepathy, 29, 130n7, 131n16, 133n1
ten-thousand-hour rule, 62, 64
"Things We Have Lost," 104, 117
Thornton, Davi Johnson, 16
thought, 23, 28, 39–42, 84, 107; materiality
of, 4, 129n1, 130n7, 131n16
3-D printing, 31, 133n3
Through the Wormhole (TV program), 67–68
training, 3, 13, 20–21, 110; centers for, 49,
134n3, 135n10, 135n12; discourses about,
50–51; gymnastic metaphors of, 55, 58–63;
mastery paradigms of, 66–68, 125; virtual,
63–64. *See also* accelerated learning;
biofeedback; brain training; mastery;
neurofeedback
translation, 24, 28, 30–31, 125; data, 39–46,
109–112
transparency, 27, 29, 46; aesthetics of, 20,
24–25, 27–28, 33; chic, 39

UCLA Center for Sleep Research, 73
unconscious, 12, 21; memories as, 117; shifts
in conceptions of, 50–54, 123; sleep and,
72, 80, 83, 85–87, 91
urban planning, 2, 97–103, 121

Van Der Drift, Marcel, 96–97, 103
Vidal, Fernando, 14
visibility: arousal and, 27–29; bodily
interiority and, 1–2, 31–32; brain waves
and, 4, 19, 36; mapmaking and, 98–99,
106–107; new forms of, 50–53, 63, 116,
123, 129n2
voluntary control, 12, 51–52, 135n6